AF789614

VEGETATION DEGRADATION IN CENTRAL ASIA UNDER THE IMPACT OF HUMAN ACTIVITIES

Vegetation Degradation in Central Asia under the Impact of Human Activities

by

Nikolai Kharin

Center on Problems of Ecology and Productivity of Forests
Russian Academy of Sciences, Moscow

SPRINGER SCIENCE+BUSINESS MEDIA, LLC

A C.I.P. Catalogue record for this book is available from the Library of Congress.

ISBN 978-94-010-3898-0 ISBN 978-94-010-0425-1 (eBook)
DOI 10.1007/978-94-010-0425-1

Printed on acid-free paper

All Rights Reserved
© 2002 Springer Science+Business Media New York
Originally published by Kluwer Academic Publishers in 2002
Softcover reprint of the hardcover 1st edition 2002
No part of the material protected by this copyright notice may be reproduced or utilized in any form or by any means, electronic or mechanical, including photocopying, recording or by any information storage and retrieval system, without written permission from the copyright owner.

TO AZA AND JANE

CONTENTS

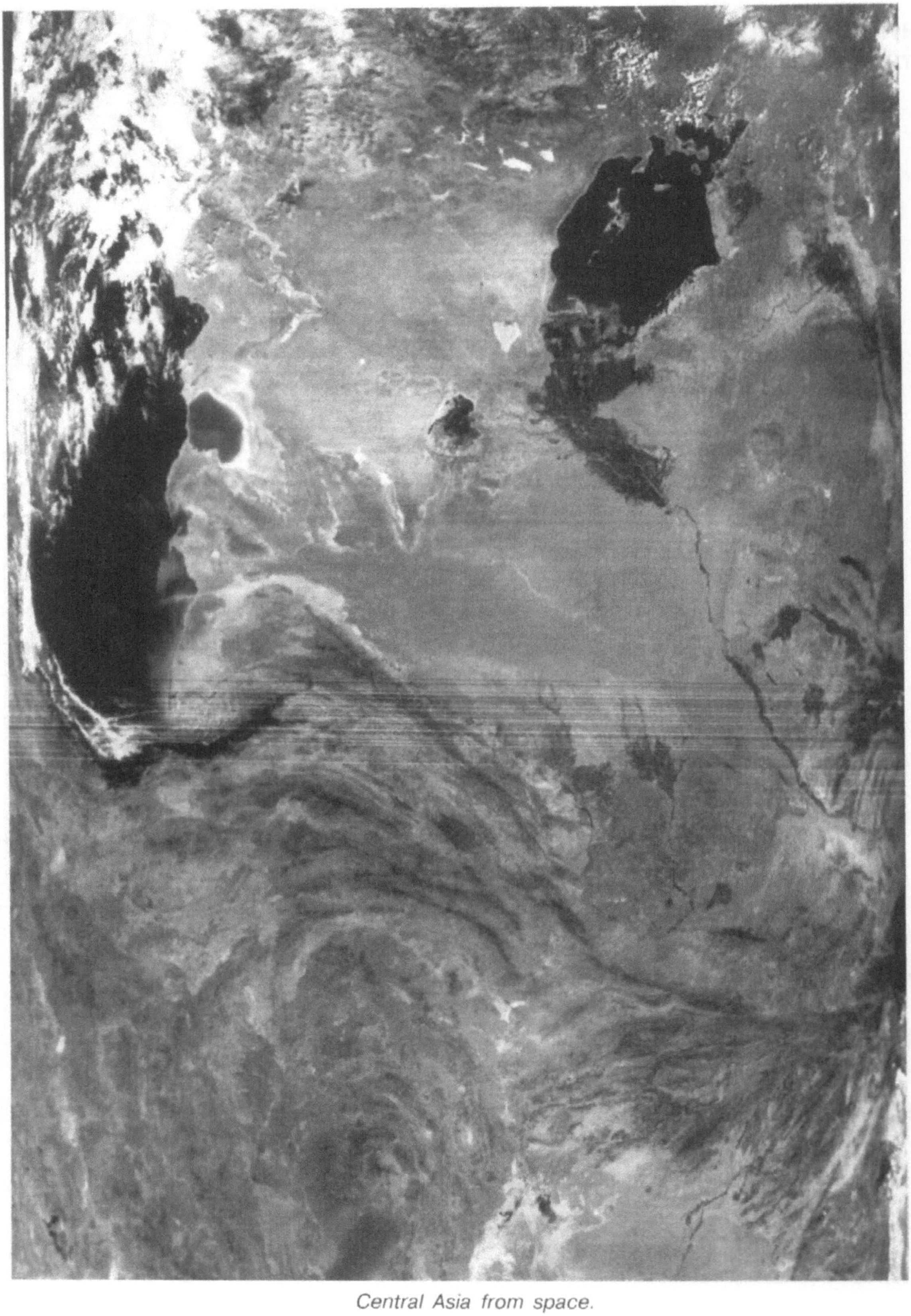

Central Asia from space.

PREFACE

Yo que en el campo he nacido
digó que mis cantos son
para los unos... sonidos,
y para otros... intención.

H. Hernandes – Martin Fierro.

The period of active influence of the people on desert environments in Central Asia counts more than 2,000 years. It includes several stages of the social and the economic development of human society – from primitive culture of hunters to developed modern society with great changes of environment and "nature remaking". Heritage of the past is one of the main causes of land degradation in Central Asia. So, only historical analysis of relations between the man and the desert could give information for awareness of desertification.

What an idea have the people about the desert? In the mind of many people "desert" is identified with a sterile, lifeless and gloom land unsuitable for life. The mass media, specially the cinema supports this idea. Very popular subject of many films is the death of travelers in the desert from thirst. Even the names of hot winds (the "gibly" in Sahara, the "samum" in Arabian desert and the "afganets" in Central Asia) fill the minds of people with horror. There are many legends about armies buried in the sand, about flourishing oases and towns covered with sand drifts. In these legends the truth is mixed with fiction. But as known, "the truth is stronger than fiction".

What is the truth? In reality the desert is a part of natural ecosystem of the Earth. The forms of biological life and ways of adaptation of living things to arid environment are numerous and often they are striking. For example, a German expedition visited Sahara in 1942. In the desert Germans found snakes – the only living things. Neither animals nor plants existed in this place. As scientists discovered later, the snakes fed on birds which took a rest on rocks in the night. The role of drylands in history of mankind is remarkable. The most ancient human civilizations – Sumar, Babylon and Assyria were located in the desert zone. And probably, the Eden was the first oasis in the desert!..

Crop production and livestock raising were the main occupations of the people in the deserts. Dry climate and droughts were not obstacles for development of the human society. Desert gave food, clothing and dwelling to the people. Desert inhabitants always loved their homeland. The Mongols living in the Gobi desert say that their country is the most beautiful in the world, water in Mongolia is the purest, and the sky is the most blue! The Kazakhs are also characteristic dwellers of deserts and steppes. They love the desert and the view of built-up land or forests causes the feeling of discomfort for them. The people who traveled in Central Asia could see the Kazakhs usually working on small railway stations in the desert.

Many dwellers of the desert after visiting other countries come back to their houses where they spend the rest of life.The Karakalpaks prefer to live and to die on the land of their ancestors, even if the ecological situation is intolerable. Their love and affection for the motherland are striking!

Desert inhabitants have adapted themselves to severe arid environment. They formed a specific diet and regime of work and recovery of their health. They invented tonics, like koumiss and chal, the last being prepared from camel milk in Turkmenistan. As regards the green tea in the life of nomads, this drink is worth to be glorified by bards.

People living without refrigerators can preserve food. For example, the people of Central Asia prepare kaurma (mutton roasted in fat). Kaurma can be conserved for a long time. Dried fruits - currants and kuraga (dried apricots) – prepared by skilful hands of Uzbek women are well known beyond the limits of Central Asia. Dwellers of the desert have created an original culture and they are proud of it. Many hand-made goods created by artisans have an artistic value. For example, Turkmen carpets decorate the houses of people in many countries.

The technological revolution of the 20th century has transformed desert lands. Railroads and highways have crossed vast desert areas. Exploitation of mineral resources, especially oil and gas, has changed the mode of people's life. Some countries, like Kuwait and Saudi Arabia, have enriched themselves on the account of their natural resources. They improved living conditions of the people. Other countries, like Central Asian ones, suffer from poverty and ecological crises. So the people can transform the desert into a flourishing land or can destroy the environment.

The process of land degradation is not just a recent. Its age is equal to the age of human civilization. The decay of many ancient civilizations was caused by soil salinization, cutting forests and overgrazing. This process has been continuing for many centuries. The well known words that " The Ottoman Empire had been eaten up by goats" could confirm that. Therefore, desertification should be studied retrospectively, embracing the whole period of human impact on the environment. Naturally, the deeper look we take into antiquity, the less information we could find. But investigations of historians and archeologists can give valuable information on development of human society. Modern scientific methodology, especially remote sensing can be used for reconstruction of the actual situation peculiar to ancient landscapes.

Desertification is a complex ecological problem, but degradation of the vegetative cover is the most pronounced indicator of this process. So, the author has paid more attention to analysis of this phenomenon. As a rule, during an historical process, cutting vegetation for fuel and domestic use as well as overgrazing could bring misfortune to the inhabitants of deserts. Social and economic consequences of vegetation degradation often were catastrophic for human society.

Following this approach, the author has studied the whole period of active interference of man in the desert environment in Central Asia. Political regimes on the territory of Central Asia changed several times during the last century. The Russian Empire conquered the khanates of Central Asia in the late 19th century. Central Asia was a part of the USSR during 74 years (1917 – 1991). New independent countries appeared here in 1991. Drying up of the Aral Sea could be considered as beginning of a global ecological crisis. The impetus to development of severe land degradation in Central Asia was given during the Soviet period when political leaders of the USSR decreed the construction of the Karakum Canal and development of new lands. According to political ambitions of Soviet leaders, all these measures should improve the prosperity of the Soviet people by transition to new society: communism. But as known, "The path to the hell is paved by good intentions"...

New independent countries of Central Asia have inherited the problem of desertifiation from the former USSR. But political instability in the region and poverty are the main bottlenecks in desertification control. In spite of that, these countries have started new efforts on rehabilitation of degraded lands.

Who is responsible for an ecological catastrophe in Central Asia? Who is guilty? Who is to blame for desertification? "Bad ecologists" from the higher echelon of power or the plain people, who voluntarily or not supported the policy of "nature remaking?" Under the present situation only public opinion can pass the verdict of guilty to those people who committed the ecological crimes. A new society, a new civilization, is created on the ruins of the Soviet empire. The new society in Central Asia can survive only if the local nations will pay their debts to mother nature, the benefactress of all peoples on the earth. A healthy society could exist only in a healthy environment.

Nikolai Kharin

February 12, 2000, Moscow

ACKNOWLEDGEMENTS

I worked during more than 30 years in the Desert Research Institute, Ashkhabad, Turkmenistan. During this period I had the opportunity to travel to many places in Central Asia, especially in Turkmenistan. Many persons helped me in my field study in the Karakum Desert and in other regions of Central Asia. The administration of the Desert Research Institute, and first of all A. Babaev assisted me in conducting these field missions.

My former post-graduates and young scientists should be mentioned here because they made a valuable contribution to collection of the field data. I want to express my thanks to T. Babaeva, B. Ivakhov, P. Radziminsky and V. Volovik.

I also want to emphasize my fruitful cooperation with the scientists of the elder generation, especially with M. Petrov, N. Nechaeva and B. Kirsta. Discussions with them and exchange of ideas helped me to develop my own concept in assessment of desertification.

During the last years of my activity in the Desert Research Institute (1992 – 1996), I participated together with A. Babaev in the development of a National Action Program to Combat Desertification in Turkmenistan. Development of this project was possible only in cooperation with some scientists, so I want to mention the names of G. Kalenov, A. Kiril'tseva, A. Batyrov, A. Ovezliev, I. Svintsov, P. Esenov and O. Kurbanov.

I also want to mention the financial and methodological support in implementation of projects on desertification control given from UNEP. In connection with that I want to express my deep gratitude to L. Kroumkatchev, T. Darnhofer and A. Vavilov.

My deep thanks to Dr. Ryutaro Tateishi from Chiba University, Japan. He helped me to compile a desertification map of Asia and to study desertification in Central Asia by use of remote sensing.

Next, several persons helped me to write this book. They discussed with me some parts of the manuscript. Thank you for your advice: P. Gunin, I. Gringof and A. Prishchepa who submitted several photos for publication in this book.

Prof. H. E. Dregne, Texas Tech University, USA, is gratefully acknowledged for scrupulous editing of the English text of this monograph.

CHAPTER 1

INTRODUCTION

Es zog einmal eine grosse Karawane durch die Wüste.., wo man nichts als Sand und Himmel sieht.

W. Hauff – Die Karawane.

1.1. United Nations Convention to Combat Desertification (CCD)

What is desertification? The United Nations Convention to Combat Desertification (1994) defines desertification *"as land degradation in arid, semiarid and dry sub-humid areas resulting from various factors, including climatic variations and human activities"*. Desertification (land degradation) includes: soil erosion caused by wind and/or by water, deterioration of the physical, chemical and biological or economic properties of soil, and long-term loss of natural vegetation.

CCD was developed in 1994 and more than 150 countries signed it (except such countries like USA, Russia and Australia). In 1997 CCD came into effect. CCD acknowledges *de jure* that "desertification and droughts are problems of global dimension", but *de facto* desertification is a global problem. Rich countries didn't want to include into CCD this definition because in this case they had to create a global fund to combat desertification.

CCD emphasizes the importance of the problem for African countries, as the title of CCD said. That is the result of activities of lobbyists for African countries. This situation is typical for activities of international organizations, because the problem of desertification has been transformed from an ecological to a political problem. As the last investigations have shown (Kharin, Tateishi & Harahsheh, 1999), the ecological situation in several Asian countries is not better than in Africa (See Section 4 of this Chapter). For example, in India the population doubled during 1951 –1991, and half of the people live under conditions of absolute poverty. The situation in Pakistan and in Afghanistan is not better. In China, severe desertification was registered on 43.4% of the drylands of the country, and the lands of the most severe type of desertification occupied 19.7%. The ecological situation in loess plateaus of China can be characterized as catastrophic. Natural vegetation in this region covers only 10% of the total area. Since 1949, the eroded land increased by 30%. Winds carry so much dust that villagers often have lunch by the light of lanterns. The living standard of peasants is very low, being only $20 per year per person!..

But we also want to indicate the importance of the problem for Central Asia countries because CCD has the lines: "Expressing concerns over impact of desertification and drought on affected countries of Central Asia and Transcaucasus".

The importance of CCD for the whole world is absolutely evident in spite of all shortcomings.

CCD not only gives definition of desertification but also defines the terrestrial object of desertification assessment – the land. Land means the terrestrial bio-productive system that comprises soil, vegetation, other biota, and the ecological and hydrological processes. It is the end of long discussion on the subject: "What is desertification?" Desertification is neither degradation of desert landscapes, nor degradation of arid ecosystem, nor desert vegetation, etc.

1.2. Global extension of desertification

Arid lands are defined as those with a net annual moisture deficiency at the surface and may be delineated by the climate criteria of Köppen (1918). If calculated by these criteria arid lands cover 42.8 mln. km^2. That makes roughly about one third of the Earth's land surface, and contain 9% of the world's population (Eyre,1988).

In "The World Atlas of Desertification" (1992) boundaries of arid lands are delineated by the Aridity Index *(AI)* calculated by the equation:

$AI = P/PET$,

where:

P – precipitations, in millimeters

PET – potential evapotranspiration, in millimeters.

The zonation of arid lands by *AI* was done in the Atlas by the following criteria:

- Hyper arid lands ($P/PET<0.05$)
- Arid lands ($0.05<P/PET<0.20$)
- Semiarid lands ($0.20<P/PET<0.50$)
- Dry sub-humid lands ($0.50<P/PET<0.65$)

Table 1.1 gives distribution of world's arid lands by regions.

These areas differ in size from that estimated by Koppen's methodology. Certainly these global figures give only a rough estimation of the area of the world's drylands. Scientists working in different arid regions estimate the areas of drylands on the basis of a complex approach. They take into consideration not only climatic features but also other environmental criteria. Petrov (1973) said that AI could be considered as rough criteria. In his opinion, it couldn't be used for delineation of arid regions. But *AI* could be used to compare dryness of climate in different areas.

According to UNEP, arid lands are the habitat and the source of livelihood for about one-quarter of the world's population.The international community is deeply concerned with the problem of land degradation occurring in all arid regions of the world. Every year nearly six mln. hectares of lands in arid regions loss their productivity. This destruction costs the world some $42.3 billions (World Atlas of Desertification,1992).

Desertification undermines the biological productivity and economic potential of fertile lands and the people at last suffer from poverty. The main causes of desertification are shown in Table 1.2.

But this percentage doesn't show the importance of each cause. For example, economic loss from salinization registered on comparatively small area could be greater than by overstocking.

Population growth stimulates the progress in desertification. Eyre (1985) identified 5 types of socio-ecological systems in arid regions of the world (Table 1.3).

Only 5% of population is nomads and semi-nomads. But this socio-ecological system occupies a considerable part of the areas in many arid regions, including Central Asia.

According to Cardy (1994), the gap between the rich and the poor in the world continues to expand. In 1960 20% of the population of the world had income thirty times greater than the poorest 20%.By 1990 the richest 20% had income 60 times greater than the poorest 20%. In 1988 10 mln. people could be considered environmental refugees. By projection 150 mln. environmental refugees from all causes will be registered by 2050.

Information on extent and severity of desertification in different continents is given in Table 1.4.

Up to-date situation

In the latest epochs several other species of plants and animal disappeared. The present situation is characterized in publication of Bie & Imevbore (1995). Information on the change

Table 1.1. Distribution of world's arid lands, in mln. ha (World Atlas of Desertification, 1992).

Total	Zones	Africa	Asia	Australia	Europe	North America	South America
1294.7	Dry sub-humid	268.7	352.7	51.3	183.5	231.5	207.5
2305.3	Semi-arid	513.8	693.4	309.0	105.2	419.4	264.2
1569.2	Arid	503.5	625.0	303.0	11.0	81.5	44.5
	Hyperarid	627.0	277.3	0.0	3.1	25.7	978.1
4852.6	Total	1958.0	1949.1	663.3	299.7	735.5	541.7

Table 1.2. Causes of desertification, in percentage (Le Houerou, 1994).

	Overcultivation	Over stoking	Fuel & wood colllection	Salinization	Urbanization	Other
N.-W. China	45	18	18	2	3	14
N. Africa & Near East	50	26	21	2	-	-
Sahel & East Africa	25	65	10	-	-	-
Central Asia	10	62	-	9	10	9
USA	22	73	-	5	?	-
Australia	20	75	-	2	1	-

Table 1.3. Socio-ecological systems in arid regions of the world (Eyre, 1988)

Socio-ecological systems	1960		1985	
	Population, mln.	%	Population, mln.	%
Nomadic and semi-nomadic stock raising	20.5	9	19.0	5
Dry and irrigation agriculture	134.0	59	256.0	62
Exploitation of mineral resources and industry	9.5	4	18.0	4
Tourism and recreation	3.0	1	6.5	2
Light industry and services	60.5	27	112.5	27
Total	**227.5**	**100**	**412.0**	**100**

Table 1.4. Extension and severity of desertification (Le Houerou, 1994).

Region	Light* S/%	Moderate S/%	Strong S/%	Severe S/%	Total**
Africa	1180/9	1272/10	707/10	35/0.2	12860/55
Asia	1567/9	1701/10	430/3	5/0.1	16718/42
Australasia	836/13	24/4	11/0.2	4/0.1	6633/13
Europe	138/5	807/27	18/0.6	31/1	2997/32
N. America	134/2	588/8	73/0.1	0/0	7324/50
S. America	418/8	311/6	62/1.2	0/0	5160
Total	4273/8	4703/9	1301/2.5	75/0.1	5169/39

* S - area desertified in 1000 km^2, % - area desertified in percentage of total drylands,

** S - total drylands, % - percentage of desertified areas in non-hyperarid lands, i.e. excluding the true desert ($P<50$ mm/year).

1.3. Change of biological diversity

Desertification being a destructive process hits the biological diversity of arid ecosystems everywhere in the world. Species composition within habitats (alpha diversity), between habitats (beta diversity) and between regions (gamma diversity) are reduced under the impact of desertification. Many species of plants and animals have been lost for always.

Silphium

Africa has suffered especially seriously during the last decades. Many major mammals have largely disappeared. Genetic diversity of many wild food and medical plants is greatly reduced. Some of them disappeared already in ancient times, for example *Silphium*. A legend said that the last specimen of Silphium had been presented to emperor Nero (Kharin & Pratov, 1982).

Silphium (also known as Laser or Laserpitium), one of the most mysterious plants of the ancient world, decorated the State emblem of Cirena town in the 7th century B.C., now it is the territory of Libya (Soderling-Brydolf, 1973). Two images of *Silphium* are known to archeologists and historians (Figure 1.1). Description of this plant made by ancient scientists is also known. Ancient sources said that *Silphium* bore resemblance to ferula and its leaves were like to leaves of parsley. It was a herb-like large plant and its roots, stalk and leaves were eatable and had medical properties. Some botanists of the modern time identify *Silphium* as *Tapsia garganica*, a plant from the Libyan flora. According to H. Keith (1965) *Silphium* variously was identified by several scientists as:

Ferula foetida, Umbelliferae,
Angelica archengelica, Umbelliferae,
Angelica officinalis, Umbelliferae,
Salvia salvatica, Labiatea,
Daucus sp., Umbelliferae.

But the properties of *Silphium*, as described in ancient literature, do not appear to fit any of these plants. According to a special study conducted by Kharin & Pratov (1982) in Libya, *Tapsia garganica* is not *Silphium* because it is a poisonous plant having morphological features different from the ancient descriptions of *Silphium*. According to their hypothesis *Silphium* is an extinct plant species. Probably it be longed to *Tapsia* genera, but it is not *Tapsia garganica*.

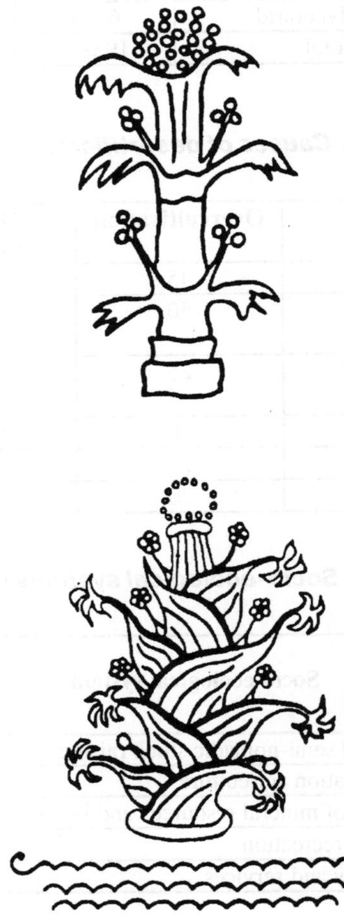

Figure 1.1. Picture of Silphium *on the medals (upper picture) and on the coins of Cyrena (lower picture), published by Keith (1965) and by Kharin & Pratov (1982).*

of biodiversity in different counties of the world is given in Table 1.5.

As known, the drylands have many limitations for the people's life. Desertification forms a threat to the biological diversity, that can remove the livelihoods of people. The losses of animal and plant species or microorganisms can't be replaced, we lose them for ever. Sustainable development in drylands can be achieved only with conservation of biological diversity, especially taking into consideration the possibility of the access to genetic material. So, the Convention to Combat Desertification calls all counties to develop joint actions in cooperation with other conventions, in particular with the Convetion on Biological Diversity.

1.4. Degradation of the drylands of Asia

Classification of the drylands of Asia

Information giving in Table 1.4 is rather rough because of different sources were used for its preparation. As to the territory of Asia, we have more correct information published by Kharin, Tateishi & Harahsheh (1999). These authors compiled a desertification map on the basis of NOAA/AVHRR data. These date were found to be similarly correlated with so called NDVI (Normalized Differences Vegetation Index), 1 km resolution (See Figure 1.2). Above that, the territory of Asia was divided in the following categories of aridity according to the landscape concept of Russian geographers (Kharin, Kalenov & Volovik, 1993):

- Semi-arid lands, low and high plains with precipitation 250 – 400 mm/year, with semi-desert shallow soils and semi-desert vegetation;

- Arid lands, low and high plains with precipitation 250 – 50 mm/year, with desert soils and desert vegetation;

- Extra-arid lands, low and high plains with precipitation less than 50 mm/year with fragments of primitive soils and sparse desert vegetation.

According to Petrov (1973), such an approach in arid land regionalization is the most objective. He indicated the following types of boundaries by division of dry lands: climatic, lithoedaphic and landscape. The essence of boundaries is unequal, namely:

- The main characteristic of the climatic boundaries is their diffusion and the dynamics in connection with climatic variations;
- The boundaries of lithoedaphic desert types are more definite and stable, and they are often linear;
- The landscape boundaries in many cases coincide with lithoedaphic, geomorphological and geobotanical boundaries, they have more complicated character, and sometime they are linear.

The landscape boundaries of desert are stable. They don't fluctuate during hundreds and thousands years (if the man doesn't interfere). Extra-arid lands were excluded from assessment because a low level of human activities.

The authors of publication mentioned above calculated the size of the drylands of Asia (Table1.6). Table 1.7 indicates the lands excluded from assessment because the human activities here didn't take place.

From these Tables we can conclude the following. The results of desertification mapping by type of land degradation show that vegetation degradation in Asia, with 58.97% of the total area, occupies a high rank of importance. Wind erosion becomes second in rank of importance with 25.4% of the total area. Water erosion and salinization have approximately the same level of importance with 7.73 and 7.18% respectively.

Desertification of the drylands of Asia

Table 1.8 gives information on land degradation in Asia in absolute figures (in km^2), and Table 1.9 gives the same information in percentage.

Social sequences of desertification

In conclusion we want to emphasize several specific features of desertification in the countries of Asia.

The countries of the Middle East have their own specific problems. As known, irrigation agriculture was practiced here four or five millennia ago. At present time, degradation of rangeland and soil salinization are the urgent problems of these countries. Rich countries, like Saudi Arabia and Kuwait can reclaim land for agricultural production. In case of need they can purchase food abroad. But poor countries, like Jordan and Yemen, suffer from desertification.

China has also a long history of land reclamation. Since the 12th century poverty in China increased because of population growth. This process is being continued in spite of measures

Table 1.5. Change of biodiversity in the drylands of the world (Bie & Imevbore, 1995)

Main causes of biodiversity changes	**Change of biodiversity**	
	Plants	**Animals**
1	2	3
North Africa Clearing of land for cultivation, overgrazing, excessive collection and uprooting of woody plants	Several plant species disappeared during the last years, about 100 exotic plants are naturalized	The following species are on the verge of extinction: the cheetah, the leopard, the striped hyena, the mountain gazelle, the barbary sheep, the baboon
Sahel Clearing of land for cultivation, heavy grazing pressure, cutting of wood, shifting cultivation	Acacia species, *Balanites aegiptica, Prosopis africana, Schoenefeldia gracilis* and other plants are replaced by weeds	Many major mammals have disappeared: the leopard, the cheetah, the elephant, the Nile crocodile, the giraffe, the West African chimpanzee
Countries around the Horn of Africa Drought, dry farming, overgrazing, cutting wood, human pressure	Valuable perennial species, like *Themeda, Cynodon dactylon, Cenchrus ciliaris* and *Digitria macrobelfera* are replaced by weeds (Sudan). In Somalia threatened species include: *Eupforbia cameroni, Wissmaria caririensis* and *Cardeauxia edulis*	Several mammals, birds and insects are on the verge of annihilation. In Uganda three species of antelopes – the bongo, the bay duiker and the white-bellied duiker – are extinct. The giant eland and the steenbok are probably extinct. The elephant and the black rhinoceros are severely reduced in number
South Africa Drought, heavy grazing and soil trampling by livestock, agricultural land development, deforestation, poaching, tourism	Deep rooted woody species are replaced by shallow rooted short living plants	Several animals, in particular big animals are reduced in number
South Europe Agricultural use of land, increase of population, urbanization, industrial development, tourism. Increase of forest areas during the last years can be considered as a positive factor of biodiversity conservation	No information	The wild boar, the lynx, the wolf and the forest cat are endangered species
Central Asia Overgrazing, cutting trees and shrubs for fuel, human pressure, agricultural encroaching, construction of the Karakum Canal, poaching	During the last century three plant species were lost: *Mandragora turcomanica, Homolodiscus ochradeny* and *Sorbus graeca*. Several other species are on the verge of annihilation	The most endangered species are: *Gasella subguttosa* and *Ovis ammon*. The tiger has disappeared. Population of snakes is reduced (*Naja oxiana* and *Vipera libetiana*). Two species of fish and one species of frog were introduced in the Karakum Canal

Table 1.5 (continued)

1	2	3
Drylands of China		
Deforestation, agricultural development, population growth, collection of medical plants	25 plants belong to rare and endangered species, in particular: *Prunus mongolica, Ferula sinkiangensis, Astragalus membranaceum,* etc.	The wild horse is in danger of extinction. The list of endangered animals includes 32 species, in particular: *Panthera unica, Grus grus,* etc.
Indian arid regions		
Cutting trees and shrubs for fuel, loss of habitats for grassland species, salinization of soils, waterlogging	47 species are known to be endangered: *Abutilon bidentatum, Alysicarpus monilifera,Anticharis glandulosa, Ephedra foliata,* etc.	11 animals and 6 birds belong to endangered species. In particular, to them belong: the wolf, the desert fox, the desert hare and the gazelle.
Australia		
Grazing by domestic livestock, seasonal fires lit by aborigines, large wild fires, introduction of exotic animals, especially the rabbit, the fox and the cat	8 species are endangered, 6 species of the arid regions have become extinct	12% of the arid zone mammals have become extinct, 85 of birds are endangered
North America		
Agricultural development, reduction of repaired areas, population pressure, urbanization, hunting and poaching	Abundance of many valuable species has been reduced. Several exotic plants are introduced, e.g. *Tamarix* and *Eragrostis* species	Several animal species have been extirpated, e.g. the California desert turtle, the buffalo, and the beaver
Latin America		
Collection of wood for fuel, agricultural reclamation of new lands, collection of medical plants, population pressure, tourism	Resources of many species are scarce, in particular *Parthenium argentatum, Prosopis sp., Acacia sp., Balsamocarpon brevifolium, Opuntia ficus-indica*	No information

taken. A new project on water transfer from south to north is discussed in China. In case of realization of this project, a new ecological catastrophe will burst in China.

Afghanistan is a country where a new type of desertification – war desertification – has destroyed the infrastructure. The war is continuing, so the rehabilitation of productive land is impossible.

Desertification in India threatens the living conditions of the people in rural areas where 46.1% of the population live in poverty. The situation in Pakistan is not better. Military clashes between these two countries took place in May 1999. That could only aggravate the ecological situation in Jammu and Kashmir provinces.

Mongolia was a satellite of the USSR for 70 years. The equilibrium between the nomadic society and deserts was destroyed during the last decades. These pastoral communities can survive only if they rehabilitate the traditional range management system.

This general information on the world status of desertification will help readers to understand better the real situation in Central Asia. Ecological catastrophe in Central Asia is not only a regional problem. CCD says that ..."desertification and drought are problems of global dimension,

Table 1.6. The area of the drylands of Asia (Kharin, Tateishi & Harahsheh, 1999).

Class names	Area, km²	Percentage
Semi-arid	3,040,189	25.49
Arid	7,294,219	61.14
Extra-arid	1,194,563	10.01
Mountains (within the drylands)	401,148	3.36
Total	**11,930,119**	**100.00**

Table 1.7. The lands excluded from assessment of desertification in Asia (Kharin, Tateishi & Harahsheh, 1999).

Class name	Area, km²	%
Bogs	514	0.02
Solonchak	121,568	5.72
Bare land	113,091	5.32
Stone surface	4,249	0.20
Mountains	401,148	18.87
Moving sands	290,268	13.66
Extra-arid lands	1,194,563	56.20
Total	2,125,419	100.00

Table 1.8. Desertification of land use types of Asia, km² (Kharin, Tateishi & Harahsheh, 1999).

Land use / Land degradation type	Land degradation areas, km²				Percentage	
	Slight	Moderate	Severe	Subtotal	Subtotal, %	Total, %
Forest, woodland / Vegetation degradation	0	61,765	257,333	319,098	78.13	3.25
Forest , woodland/ Water erosion	6,041	32,198	12,521	50,760	12.43	0.52
Forest, woodland/ Wind erosion	0	38,563	0	38,563	9.44	0.39
Subtotal	**6,041**	**132,526**	**269,854**	**408,421**	**100**	**4.17**
Rangeland & meadow/ Vegetation degradation	1,636,595	2,634,915	1,191,224	5,462,734	68.31	55.72
Rangeland & meadow / Water erosion	86,021	321,795	0	407,816	5.10	4.16
Rangeland & meadow / Wind erosion	190,582	1,026,945	861,628	2,079,155	26.00	21.21
Rangeland & meadow / waterlogging	7,363	27,206	12,258	46,827	0.59	0.48
Subtotal	1,920,561	4,010,861	2,065,110	7,996,532	100.00	81.56
Dry agriculture/ Water erosion	32,405	251,114	16,232	299,751	44.36	3.05
Dry agriculture/ Wind erosion	216,432	155,130	4,477	376,039	55.64	3.84
Subtotal	**248,837**	**406,244**	**20,709**	**675,790**	**100.00**	**6.89**
Irrigation agriculture/ Salinization	191,697	337,378	175,240	704,315	100.00	7.18
Dried up sea floor/ Salinization	0	10.335	9,307	19,642	100.0	0.20
Total	**2,367,136**	**4,897,344**	**2,540,220**	**9,804,700**	**0**	**100.00**

Table 1.9. Desertification of the drylands of Asia, in percentage (Kharin, Tateishi & Harahsheh, 1999).

Land use/ land degradation type	Land degradation areas in percentage				Percentage of total
	Slight	Moderate	Severe	Subtotal	
Forest / Vegetation degradation	0	15.12	63.01	78.13	3.25
Forest / Water erosion	1.48	7.88	3.07	12.43	0.52
Forest / Wind erosion	0	9.44	0.00	9.44	0.39
Subtotal	**1.48**	**32.45**	**66.07**	**100.00**	**4.17**
Rangeland / Vegetation degradation	20.59	33.14	14.98	68.72	55.72
Rangeland / Water erosion	1.08	4.05	0.00	5.13	4.16
Rangeland / Wind erosion	2.40	12.92	10.84	26.15	21.21
Rangeland / Waterlogging	0.09	0.34	0.15	0.59	0.48
Subtotal	24.16	50.45	25.98	100.00	81.56
Dry agriculture/ Water erosion	4.80	37.16	2.40	44.36	3.05
Dry agriculture/ Wind erosion	32.03	22.96	0.66	55.64	3.84
Subtotal	36.82	60.11	3.06	100.00	6.89
Irrigation agriculture/ Salinization	27.22	47.90	24.88	100.00	7.18
Dried up sea floor/ Salinization	0	52.62	47.38	100.00	0.20
Percentage of total	**24.14**	**49.95**	**25.91**	**100.00**	**100.00**

in that they affect all regions of the world and that joint action of the international community is needed to combat desertification and/or mitigate the effects of drought".

Capital punishment for ecological crimes

From the Asian countries located in the arid zone, one should be mentioned here as a country where the president and other high officials are standard-bearers of nature conservation. According to one publication (Pilyatsky, 2000), this country is considered as one of the most ecologically safe territories. That is the UAE – United Arab Emirates. During the last decades several measures on nature conservation were taken in this country. They included planting dozens of billions of trees, creation of protective areas and reserves, conservation of rare animals and birds etc. Contamination of soil, atmosphere, inland or sea water is severely punished. Capital punishment can be applied for persons responsible for radioactive contamination of the environment.

Information on nature conservation and the ecological education of the population are considered as the measures of high priority in inner policy of the country. The annual ecological festival, conducted everywhere in the country in February, is a great feast in which plain people and the leaders of the country participate.

Figure 1.2. NOOA AVHRR/ NDVI 10-day composite image of Asia. The composite period of the image is 11 - 20 May, 1992. Source: U.S. Geological Survey. Processing: CEReS, Chiba University (Kharin, Tateshi & Harahsheh, 1999)

CHAPTER 2

PHYSICAL ENVIRONMENT AND AGRICULTURAL DEVELOPMENT

Entrails of the unfathomable Asia
Like a flaring volcano
Sent to us millions off-springs,
The ardent hurricane was blowing
from steppe,
And steppe was disturbed like ocean.
The Orient was fraught with apparition
of Chinggis Khan!
At that time Russia alone
Was a sentry post of Europe,
And Russian people shed blood
About their faith...
A. N. Maikov – Klermont Cathedral
(Translated from Russian by N.G.Kharin)

2.1. Introduction

Central Asia is located within the vast Aralo-Caspian basin. Being a part of the continent of Asia, Central Asia is bordered on the south by mountain chains of the Kopetdag and the Parapamis, and on the east by piedmont plains of the Dzhungarian Alatau, the Tian-Shan and the Pamir. Deserts and semi-deserts of Central Asia are divided by the highest mountains from the tropical belt of Asia. Vast desert plains devoid of surface runoff, piedmont plains and oases of irrigated soils are the main landscapes of the region. Mountain landscapes are typical for the southern and eastern parts of this territory. Central Asia occupies an area of 4,994.4 mln. km^2. Its population totaled 51.8 mln. in 1991.

The physical limit of the region from north follows the watershed line between the Irtysh river and the Aral Sea. This region was known by Russian scientists as "Middle Asia". This term needs clarification.

Russian geographers of the 19th century used the term "Middle Asia" to include the territory of three khanates: Khiva, Bukhara and Kokand. Terent'ev (1906) proposed to include in "Middle Asia" the territory between 35 – 50° N and 70 – 98° E. The southern boundary of this territory was better identified. The west boundary ran along the coast of the Caspian Sea. The northern boundary was not clear. All this territory was also known as "Turan" or "Turkestan", that means "The land of Turks". After the Russian conquest the territory became known by a new name, "Russian Turkestan".

The English geographer Rickmers (1913) limited Turkestan to the "Duab of Turkestan" – the area between two rivers – the Amurdarya and the Syrdarya (Figure 2.1). But this idea was not supported by other scientists and now the term is not used in that sense.

Western geographers usually used the term "Central Asia" instead of "Middle Asia". After the Bol'shevik revolution in 1917, they began to use the term "Soviet Central Asia". But it should be noted that Russian geographers used the term "Central Asia" for the vast arid region which included Mongolia and the northern part of China!

After 1991, new independent countries appeared in the region of consideration: Kazakhstan, Kirghizstan, Uzbekistan, Tajikistan and Turkmenistan. Political ambitions of new leaders and their striving for higher rating gave rise to a new definition – "Middle Asia" was transformed to "Central Asia". Apologists of a new doctrine gave a justification for this definition. For example, Kadymov (1993) said: " When the people of Middle Asia look around, the first thing they had seen was the central, not peripheral geopolitical location of their countries in Asia". This new geopolitical region includes the five countries mentioned above in the limits of their political boundaries (Figure 2.2).

30% of Kazakhstan's territory is plains, 50% is elevated plains, plateaus and hills, the rest is mountains. Landscapes are changing from north to south-forest steppe, steppe, semi-desert, desert.

In Kirghizstan, 75% of the territory is mountains, half of its territory being located at 1,000-3,000 m above the sea level, one third at 3,000-4,000 m. Foothills up to 1,200 m comprise a small area.

In Uzbekistan, desert landscapes occupy about 70% of the territory. Cultivated lands are confined to river valleys and deltas.

In Tajikistan, 93% of its territory is mountains, the rest belongs to plains and valleys. In this country the ecological zones are changed in the following sequence: desert, semi-desert, forest steppe, high mountain meadows and alpine belt.

In Turkmenistan, 87% is plains with sand dunes, solonchaks and clay deserts. Irrigated lands are confined to deltas and river valleys. Mountains are of low or medium elevation.

2.2. Climate and desertification

Human-induced land degradation occurs in close correlation with climatic features which can be considered as the background of desertification.

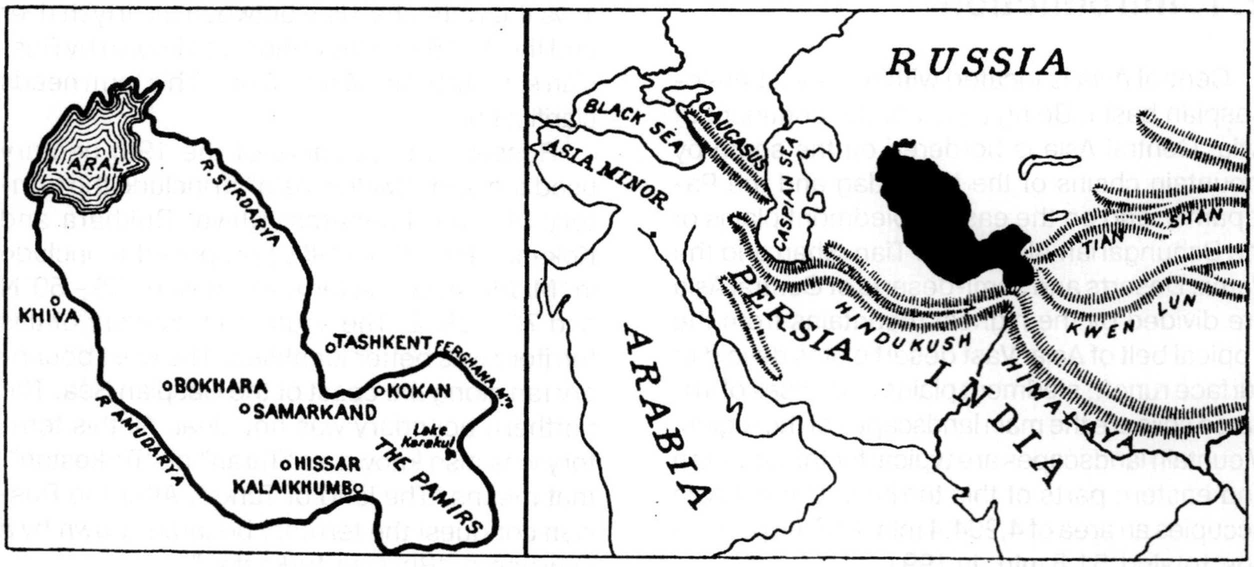

Figure 2.1. The Duab of Turkestan (Rickmers, 1913)

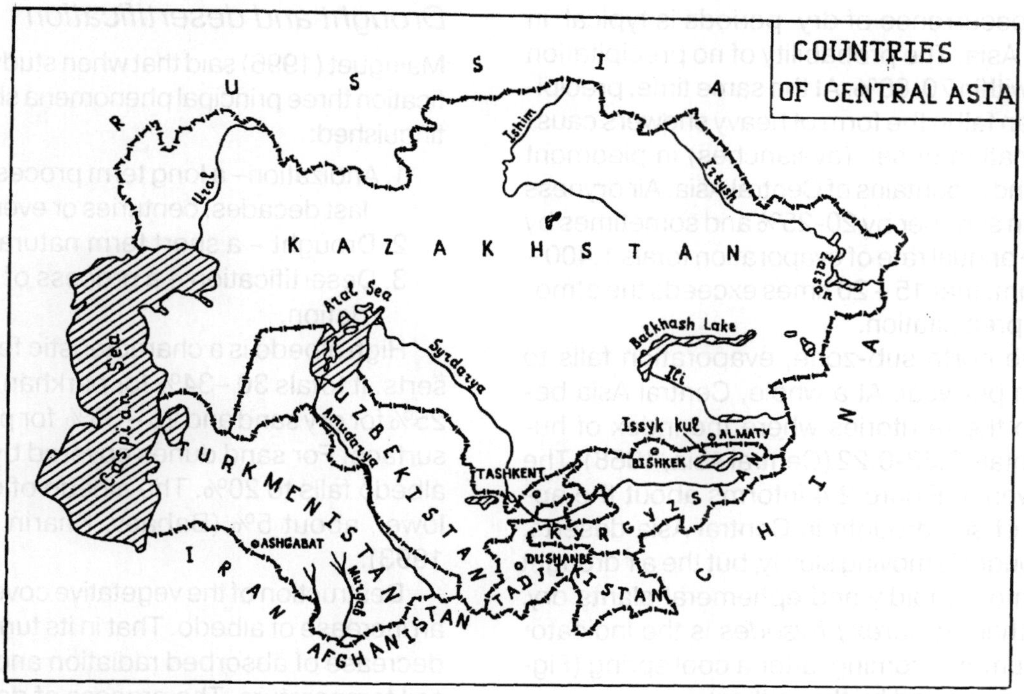

Figure 2.2. Countries of Central Asia

Specific climatic features of Central Asia

The deserts of Central Asia are located in the temporal belt of the continental climate (Petrov, 1973). Summer is hot, dry and cloudless. Winter is mild, except in the northern part of the zone where winter is cold. These climatic features are intensified by orographic barriers which isolate the territory from the south, and no such barriers are found in the north. So the temperature regime in summer doesn't differs strongly from the neighboring countries – Iran and Afghanistan. But winter in Central Asia is cooler than in these countries.

Geographers divide the desert zone of Central Asia into two sub-zones: northern and southern (Zonn, Nikolaev &. Orlovsky, 1981). Climatologists noticed that reduction in precipitation takes place sharply between 42-45° N. So they divide the two sub-zones along a line following the northern coast of Kara Bogaz Gol – delta of the Amudarya – southern skirts of the Bakantau mountains – the Karatau ridge.

Climatic criteria of the two sub-zones are given in Table 2.1, and isohyets in Figure 2.3.

Variation of annual precipitation is one of the factors defining the instability of arid ecosystems. This variation could be very great. For example, variation of the annual precipitation in Turkmenistan could range from 24 mm to 540 mm.

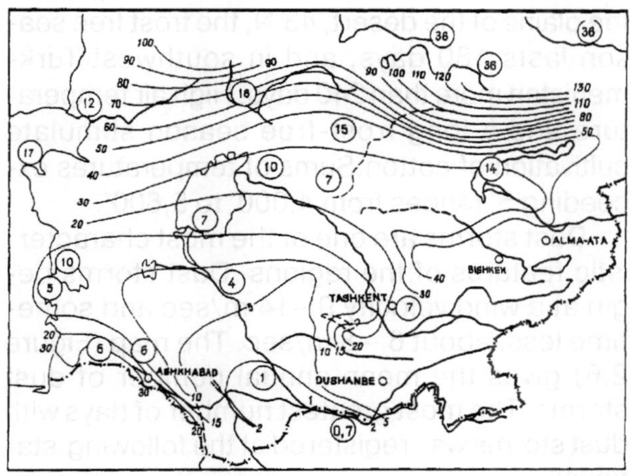

Figure 2.3. Precipitation in mm, and the number of days (in circles) with precipitations less than 0.1 mm from June to September (Zonn, Nikolaev & Orlovsky).

The occurrence of dry periods is typical in Central Asia. The probability of no precipitation ranges within 70-80%. At the same time, precipitation can fall in the form of heavy showers causing formation of sel' (avalanches) in piedmont plains and mountains of Central Asia. Air dryness can fall in summer by 20-25% and sometimes by 3%. The annual rate of evaporation totals 1,400 – 2,300 mm, that 15 – 20 times exceeds the atmospheric precipitation.

In the north sub-zone, evaporation falls to 900 mm per year. At a whole, Central Asia belongs to the territories where the index of humidity totals 0.33-0.22 (Central Asia, 1968). The map given in Figure 2.4 informs about the approach of soil drought in Central Asia deserts. This drought is moving slowly, but the air drought reacts more rapidly and ephemeral plants dry. The withering *Carex physodes* is the indicator of hot summer coming after a cool spring (Figure 2.5; compiled by the author).

The day-time amplitude of the air temperature in Central Asia ranges from 15° to 25°. So the people have a chance to rest after the heat. In the most hot summer months (July and August) the mean minimum night temperature is 17-24°.

The air temperature sharply falls at the end of September and beginning of October. The first autumn frosts occur during the second ten-day period of October in the north and at the end of October in the south parts of the region. In the plains of the desert, 43°N, the frost free season lasts 180 days, and in southwest Turkmenistan more than 240 days. High air temperature and a long frost-free season stimulate cultivation of cotton.Sums of temperatures exceeding 5° ranges from 4,000° to 5,600°.

Dust storms are one of the most characteristic features of the regions. Dust storms begin at a wind velocity 9 –14 m/sec and sometime less, about 6 – 8 m/sec. The map (Figure 2.6) gives the mean annual number of dust storms. The most greatest number of days with dust storms was registered at the following stations:

Cheshme	– 113 (1948)
Molla-Kora	– 146 (1838)
Muinak	– 121 (1958).

Drought and desertification

Mainguet (1996) said that when studying desertification three principal phenomena should be distinguished:

1. Aridization– a long term process which can last decades, centuries or even millennia.
2. Drought – a short term natural process.
3. Desertification – a process of land degradation.

High albedo is a characteristic feature of deserts. It totals 30 –34% for barkhan sands, 20 – 25% for dry sand and 25 – 35% for pebble stony surface. For sand dunes covered by vegetation albedo falls to 20%. The albedo of croplands is lower, about 5% (Babaev, Kharin & Orlovsky, 1993).

Destruction of the vegetative cover stimulates an increase of albedo. That in its turn causes the decrease of absorbed radiation and increase of soil temperature. The process of desertification could be aggravated by overgrazing. According to Otterman (1974), this process in the Negev desert can be cyclic. But Le Houerou (1992) criticized this approach. He said that this hypothesis had been never validated actually. In his opinion, the albedo feedback assumption for the Sahel was not operative.

The impact of drought and desertification on natural ecosystems is first of all marked by reduction of biological productivity. Loss of forage production on rangelands can cause catastrophic sequences in the socio-ecological systems of drylands. Le Houerou (1994) proposed a Rain-Use Efficiency criterion (RUE) for evaluation of rainfall-production correlation. According to his study, in sandy deserts the RUE value is about 4 – 6 kg/ha/year/mm of precipitation.

In Central Asia this correlation was established by An. &. Konovalova (1981). Regression equations are given in Table 2.2. If we substitute in these equations the value of precipitation of 100 mm, we may get the following RUE (per 1 mm of precipitation), in kg/ha: Akbaital – 3.1, Buzaubai – 3.5, Mashikuduk – 4.4, Kulkuduk – 4.1, Mubarek – 3.2, Zhaslyk – 4.5. By the precipitation of 150 mm RUE value for Uzbekistan may reach 4 – 6 kg DM (dry matter)/ha/year.

Table 2.1. Main climatic features of Central Asian deserts (Zonn, Nikolaev & Orlovsky, 1981).

Stations	Air temperature, degree C					Length of period with mean day temp. $<0^0$ C	Growing season, days	Sum of temperature > 10^0	Precipitation		
	Annual	January	July	Mean absolute min	Abs max				Annual	Warm season	Cool season
Teren Kuduk	4.8	-15.4	25.4	-35	43	151	157	1973	171	102	69
Ak Tumsuk	8.9	-7.9	25.2	-26	42	107	173	3666	129	77	52
Sam	8.9	-9.1	27.0	-28	46	126	179	3477	136	82	54
Kungrad	9.9	-6.7	25.2	-22	44	106	186	3571	96	46	49
Kunya-Urgench	10.9	-6.0	26.2	-24	45	102	193	4042	76	76	45
Ekedzhe	13.2	-3.9	29.7	-22	46	81	205	4697	100	40	60
Darvaza	14.8	-2.0	31.2	-19	46	56	217	3132	97	39	58
Akmolla	15.8	-1.0	32.2	-20	47	44	224	5393	91	34	57
Chezme	15.7	-0.2	31.5	-20	49	29	228	5292	97	38	59
Bairam-Ali	16.0	1.5	30.2	-16	48	0	234	5268	138	40	95
Serakhs	16.6	3.0	30.2	-15	48	0	224	5402	178	51	127
Takhta-Bazar	16.5	2.7	31.2	-17	48	0	236	5370	241	53	188

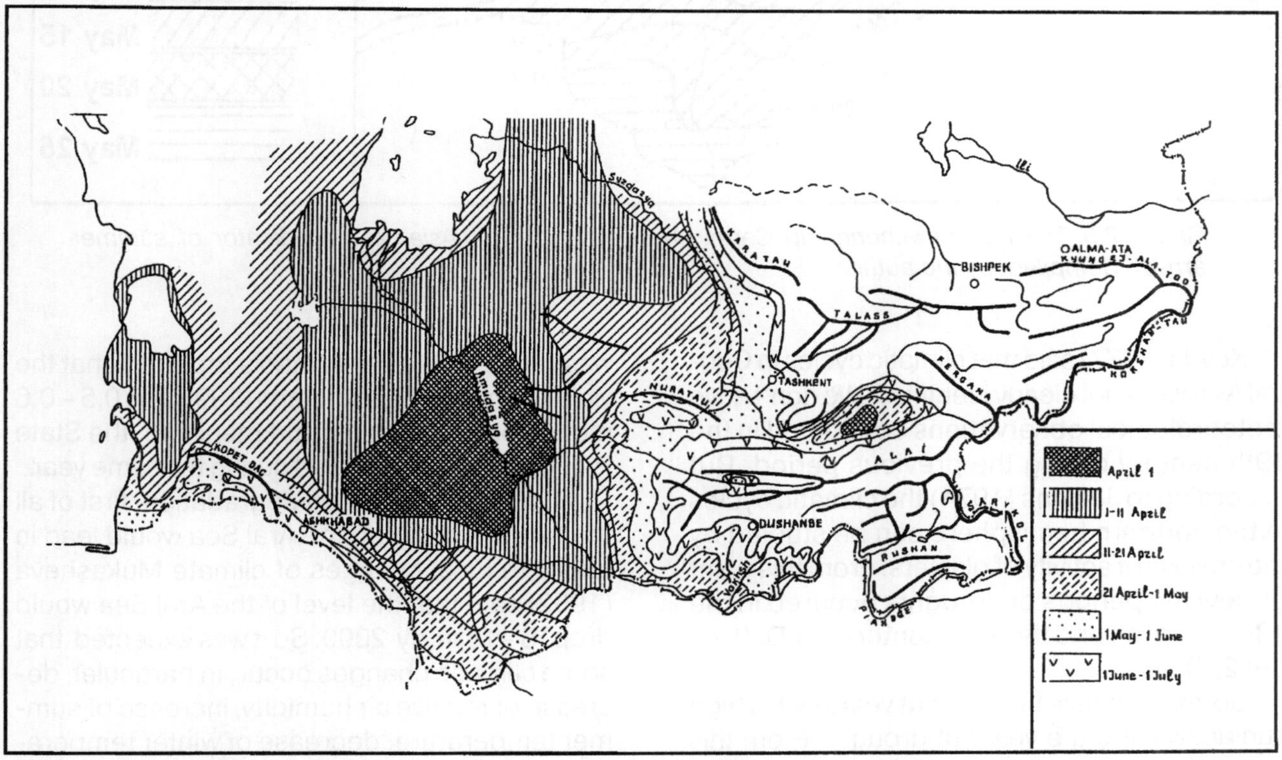

Figure 2.4. Arrival of soil drought in Central Asia (Zonn, Nikolaev & Orlrovsky, 1981)

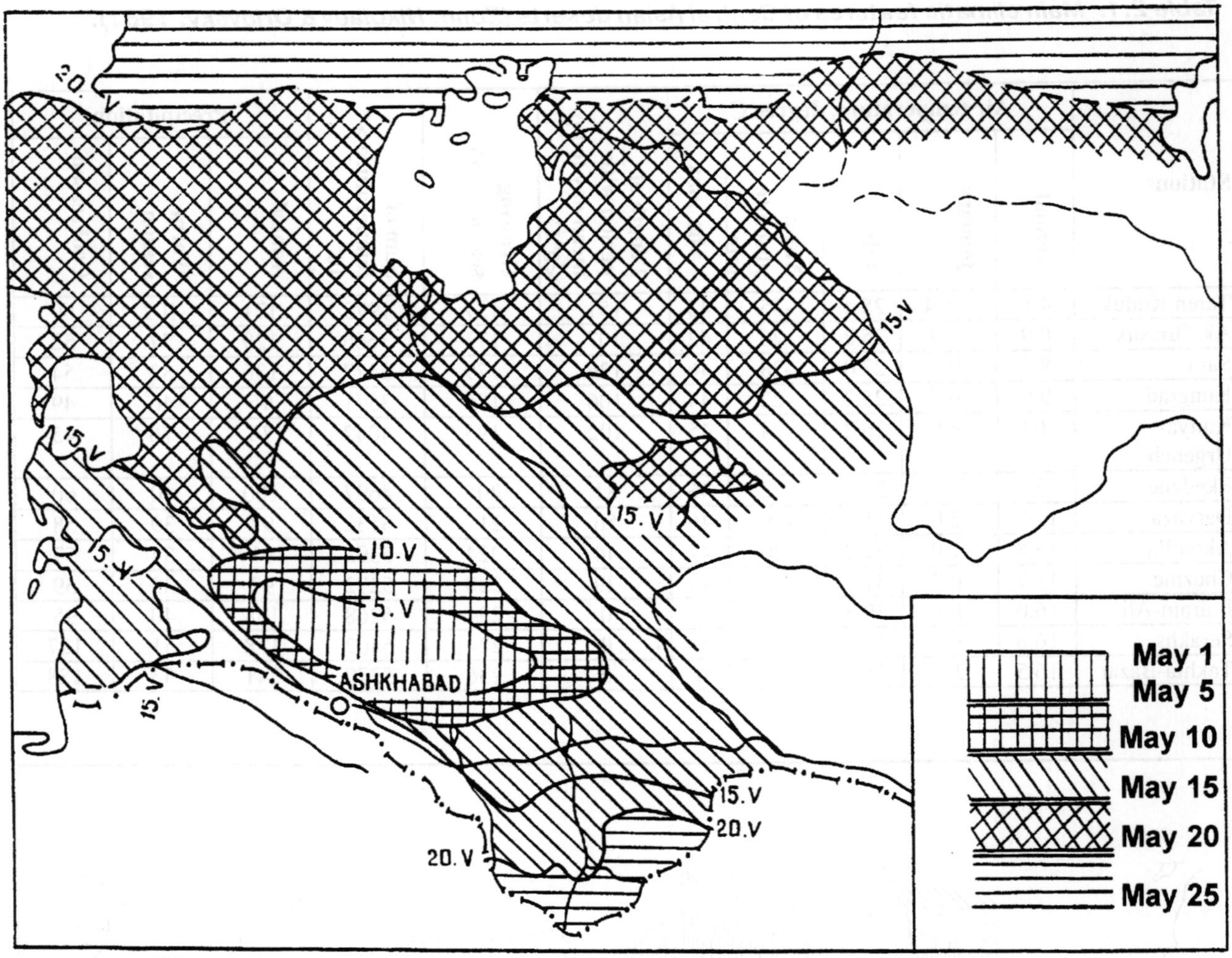

Figure 2.5. The map of withering up Carex physodes *in Central Asia, as an indicator of summer arrival. Compiled by the author.*

Kovda (1977) said that climatic cycles in Central Asia were not clearly identified. We don't have meteorological observations conducted in the 19th century during the previous period. But according to Jenkins (1974) the climatic cycles in the northern hemisphere can be studied by the change of activity of glaciers. From this point of view the periods of droughts occured in the 13th, the 15th and the 17th centuries A.D. (Figure 2.7).

Some scientists believe that years with a high sun activity are the years of drought. From this point of view, the period 1994 –1997 was expected to be dry and 1997– 2000 years would be humid. Really the years 1995 –1996 were very dry. The yield of agricultural crops during these years was very low. The mass media reported that the yield of wheat in Kazakhstan in 1995 was 0.5 – 0.6 metric tons per 1 ha. In Turkmenistan, the State Programme "Grain" failed during the same year.

Human-induced land degradation, first of all the further drying up the Aral Sea would lead in future to local changes of climate. Mukasheva (1992) said that the level of the Aral Sea would drop 16 – 18 m by 2000. So it was expected that some climatic changes occur; in particular, decrease of relative air humidity, increase of summer temperature, decrease of winter temperature, increase of number of days with dust storms etc. All these changes would lead to expansion of desert lands, formation of new solonchaks and to further aggravation of desertification.

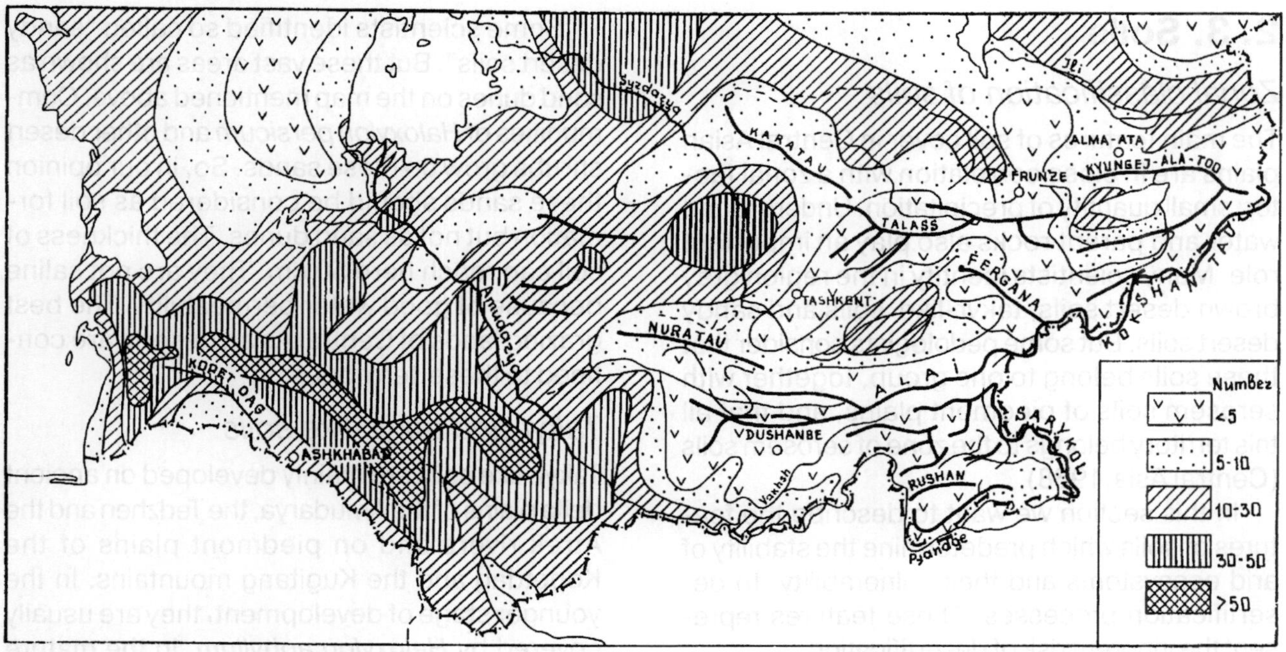

Figure 2.6. The mean number of days with dust storms per year in Central Asia (Central Asia, 1968).

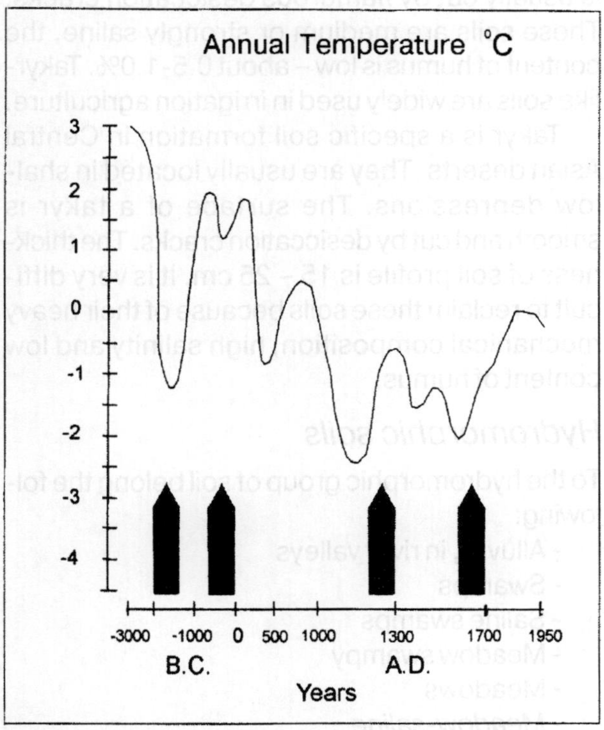

Figure 2.7. Variation of mean annual temperature in North China. Arrows show the periods of glacier activities in North hemisphere (Jenkins, 1974)

Table 2.2. Correlation between precipitation and productivity of rangelands, in center per hectare (An & Konovaliva, 1981)

Station	Equation	r	S
Akbaital	$y = 0.019x + 1.2$	0.81 ± 0.03	0.35
Buzaubai	$y = 0.026x + 0.92$	0.94	0.19
Mashikuduk	$y = 0.029x + 1.48$	0.79	0.48
Kulkuduk	$y = 0.035x + 0.15$	0.67	0.45
Dzxhangel'dy	$y = 0.035x + 0.15$	0.67	0.45
Mubarek	$y = 0.013x + 1.87$	0.76	0.34
Zhaslyk	$y = 0.030x + 1.50$	0.86	0.24

Note: y – annual forage, C /ha (C – centner, 100 kg), x – the sum of precipitation registered during 30 days before the 0 degree of air temperature threshold & precipitation registered during 30 days after the same threshold, r - correlation coefficient, S – mean annual forage, C/ha.

2. 3. Soils

Zonal classification of soils

The main features of soil cover in Central Asian plains are in close correlation with a zonal factor: small quantity of precipitation. Underground water and parent rocks also play an important role. Many scientists identify in the region gray brown desert soils, takyr-like soils and sandy desert soils. But some pedologists consider that these soils belong to one group, together with serosem soils of piedmont plains, and that all this territory belongs to the zone of serosem soils (Central Asia, 1968).

In this section we want to describe the features of soils which predetermine the stability of arid ecosystems and their vulnerability to desertification processes. These features represent the inherent risk of desertification.

According to "Soil map of Central Asian Republics" (Lobova, 1971), the following soils are common in the plains of Central Asia:

Gray brown typical
Gray brown solonetc-like
Gray brown solonachak-like
Desert takyr-like
Desert takyr-like saline
Takyr.
Desert sandy soils.

Gray brown soils are mainly common in the Usturt and on the Krasnovodsk peninsula, in the Transunguz Karakum and the south Kyzylkum, as well as on piedmont plains. Dwarf semi-shrub vegetation with low density is the most typical for these soils (*Artemisia kemrudica, A. terra-alba, Salsola gemascens, S. orientalis* and *Anabasis salsa*). According to Lavrov (1984), these soils are formed on fine sandy loam and on sandy deposits underlain by limestone and pebble diluvium. Soil material usually contains much crystal gypsum. Ground waters are usually saline, their depth is more than 20 m. The content of carbonate is very high, the content of humus is very low, about 0.5%.

In sandy deserts several types of sand deposits are met, in particular maritime sands on New Caspian deposits, sands on Quaternary alluvial deposits of Central Asia rivers and sands on Pliocene alluvial deposits.

Some scientists identified so called "sandy desert soils". But these vast areas are shown as sand dunes on the map mentioned above. Communities of *Haloxylon persicum* and other desert shrubs grow on these sands. So, in our opinion these sands should be considered as soil formation but not as sand dunes. The thickness of soil can reach here 50 cm. They are not saline because of good water permeability. The best grazing lands of Central Asian deserts are confined to these territories.

Takyr and takyr-like soils

Takyr-like soils are mainly developed on ancient delta plains of the Amudarya, the Tedzhen and the Atrek rivers and on piedmont plains of the Kopetdag and the Kugitang mountains. In the younger stage of development, they are usually covered by *Haloxylon aphyllum*. In the mature stage such species as *Salsola kemrudica, S. orientalis* and *S. gemascens* are dominant plant species. Plant cover is very low totaling 20-30%. Thickness of soil is 25 – 30 cm. The soil surface is usually cut by numerous desiccation cracks. These soils are medium or strongly saline, the content of humus is low – about 0.5-1.0%. Takyr-like soils are widely used in irrigation agriculture.

Takyr is a specific soil formation in Central Asian deserts. They are usually located in shallow depressions. The surface of a takyr is smooth and cut by desiccation cracks. The thickness of soil profile is 15 – 25 cm. It is very difficult to reclaim these soils because of their heavy mechanical composition, high salinity and low content of humus.

Hydromorphic soils

To the hydromorphic group of soil belong the following:

- Alluvial, in river valleys
- Swamps
- Saline swamps
- Meadow swampy
- Meadows
- *Meadow-saline.*

The soils of the last group are formed under permanent or temporary flooding. The degree of salinity varies from slight to severe. Tugai vegetation is common here. These soils can be reclaimed only by construction of drainage.

Soils of the irrigation zone

The area of irrigated farmlands increases from year to year. To irrigated soils belong the following:

- Previously irrigated meadows
- Previously irrigated meadow saline
- Previously irrigated desert soils
- Previously irrigated serosem
- Irrigated typical serosem
- Irrigated light serosem
- Irrigated light saline serosem
- Irrigated desert takyr-like
- Irrigated gray brown
- *Irrigated meadow.*

Near everywhere in Central Asia one may meet the ruins of ancient fortresses and traces of previous irrigation canals. The age of irrigated soils totals 2-3 thousand years in many oases of the region, especially in the lower reaches of the Amudarya, in the valley of Fergana, etc. Agro-irrigation deposits were formed as a result of long irrigation. Their thickness may total 4 m. In the process of irrigation these soils have lost their initial features: in particular the content of humus increased, the soil structure was improved and salinity fell. If plowing is stopped the soils regenerate their features. But their structure and content of humus are conserved for a long time.

Soils of vertical zonation

The soil of vertical zonality includes the following types:

- Light serosem
- Saline compact serosem
- Typical serosem
- Light serosem with low content of carbonates
- Typical serosem with low content of carbonate
- *Meadow serosem.*

Typical serosem are confined to low slopes of mountains, on elevations of 500 – 1000 m above the sea level. Vegetative cover consists of *Poa bulbosa, Carex pachystylis* and *Artemisia* species. These soils are formed by loess sandy loam or dust sandy loam. They are usually not saline and rich in phosphorus.

The group of mountain soils includes the following:

- Mountain serosem (Dark serosem)
- Mountain chestnut
- Mountain gray brown
- Mountain brown with carbonates
- *Mountain meadow alpine and other.*

Mountain serosem soils are met at the elevation of 1000 – 1800 m above sea level, with precipitation 350 mm/year (Lavrov, 1984). Common plants are *Elitrigia caespitosa* and ephemers. Mechanical composition in the upper horizon is sandy loam, becoming heavier in lower horizons. The physical features of these soils are good for plant growth.

Mountain chestnut soils are met on slopes of the Kirghiz Ridge, at an elevation of 900 – 1200 m above sea level. They contain 4 – 6% of humus. These soils are used for dry and irrigation agriculture.

Mountain brown carbonate soils are typical for low mountains in Uzbekistan and in Tajikistan, they are also met in the Kopetdag and the Kugitang mountains. They are the characteristic soils for the tall grass savanna and semi-savanna of Central Asia. The content of humus in this soil can reach 3.5%. These soils are widely used in dry agriculture.

2. 4. Vegetation

Classification of vegetation

The phytogeographical aspects of Central Asian deserts are very specific, especially concerning the plant composition. Deserts of Central Asia are centers of floristic formation of many plant families, genera and species (Korovin, 1961). Some 1600 – 1700 plant species have been registered in the flora of Central Asian deserts. The most numerous families are the following: *Chenopodiaceae* (256 species), *Compositae* (160 species), *Fabaceae* (148 species), *Cruciferae* (96 species), Liliaceae (72 species).

In the process of evolution desert plants have developed the following types of adaptation (Petrov, 1973):

1. Morphological and anatomical (extensive root system, water accumulating tissue, salt exudation glands).
2. Physiological (special features of photosynthesis and transpiration, high concentration of salts in cellular sap, etc.).
3. Biological (seasonal rhythms of plants).
4. Ecological (structure of phytocoenosis, species composition, change of seasonal aspects).

Nechaeva, Vasilevskaya & Antonova (1973) have identified the following life forms of plants in the Karakum desert: trees, shrubs, dwarf shrubs, semi-shrubs, dwarf semi-shrubs, perennial herbs, biannual plants and annuals. The seasons of growth of desert plants are indicated in Table 2.3.

The duration of the growing season in southern deserts totals 9 – 10 months, from February to December. Many plants grow 12 months during the years with a warm and long spring. In winter, mainly cold-resistance species grow; heat resistance plants, trees, shrubs and semi-shrubs grow in summer. The seasonal rhythm of plants in several cases is in close correlation with their taxonomy. For example, many *Chenopodiacea* species have, as a rule a long growing season.

Low plant cover is a characteristic feature of desert communities. Severe ecological conditions, in particular hydrology and soil salinity, are the main environmental factors effecting the structure of plant communities. The density of many plants increases under better moisture conditions. Many lower plants, e.g. mosses, form dense sinusia even in the comparatively low moisture content of desert soils.

Vegetation of sandy deserts

Vegetation of sandy deserts has the highest productivity and its plant composition is the richest. The main plants of sandy deserts are the following: *Haloxylon persicum, H. aphyllum, Calligonum sp.sp., Ammodendron conollyi, Ephedra strobilacea, Artemisia turanica, Salsola richteri, Stipagrostis pennata, Artemisia arenaria*. Saxaul in many places forms thickets, consisting of shrubs 5 – 7 m in height (Figure 2.8).

In sandy deserts, three layers are usually identified (Nechaeva & Nikolaev, 1962):

1. The first layer, formed by shrubs (*Haloxylon* species, *Salsola richteri, Calligonum species, Ephedra strobilacea*), 120 – 250 cm in height,
2. The second layer, 60 – 90 cm in height, formed by semi-shrubs and high herbs (*Artemisia kemrudica, Astragalus species, Aristiuda karelini*, etc.),
3. The third layer, 15 – 25 cm in height, formed by *Carex physodes* and annuals.

Figure 2.8. Ticket of saxaul (Haloxolon aphyllum) *in Southeast Karakum. Photo by A. Prishchepa.*

Vegetation of clay deserts

Vegetation of clay deserts is mainly formed by shrubs and semi-shrubs. The main plants are as follows: *Haloxylon aphyllum, Salsola orientalis, S. gemascens, S. arbuscula, Artemisia kemrudica, A. badkhysi, A. turcomanica, Ceratoides papposa*. Annual halophytes are also common here, in particular: *Gamacanthus gamocarpa, Halimocnemum longifolia, Halocharis hispida*. Grasses and spring ephemers are also met in these communities. Plant cover is very low.

Vegetation of gypsum deserts

The following plants of gypsum deserts can be mentioned: *Haloxylon aphyllum, Salsola gemascens, S. arbuscula, Anabasis salsa, Artemisia kemrudica, A. badkhysi, A. sublessiana, A. terra-alba, A. arenaria, Stipa sareptana, Nanopnyton erinaceum, Atriplex cana*. Only two plant layers are common in this formation:

Table 2.3. Seasons of growth of plants of different life forms, number and percentage of species (Nechaeva, Vasilevskaya & Antonova, 1973).

Life forms	All year, number / %	Winter & spring, number/ %	Spring, number / %	Spring & beginning of summer number/%	Spring & summer number/%	Spring, summer & autumn, number/%	Total number/ %
Trees		- - -	- -	- -	1/ 0.4	4/ 1.4	5/ 1.8
Shrubs	2 / 0.7	- -	- -	3 / 1.1	10/ 3.6	13/ 14.7	28/10.1
Dwarf shrubs	1 / 0.4	- -	- -	- -	- -	3 / 1.1	4/ 1.4
Semi-shrubs	1 / 0.4			6 / 2.2	4 / 1.4	3 / 1.1	14 / 5.1
Dwarf semi-shrubs	10 / 3.6	- -	- -	- -	7/ 2.5	8/ 2.4	5/ 9.0
Perennials	- -	14 / 5.1	25/ 9.0	- -	12 /14.3	9 / 3.2	60 /21.2
Biannuals	- -	- -		- -	1/ 0.4	1/ 0.4	2/ 0.8
Annuals	- -	80/ 28.8		31/11.2	15/ 5.4	13/ 4.7	139/50.2
Total	14/ 5.1	94/ 33.9	25/ 9.0	40/14.5	50/ 18.0	54/ 19.5	277/ 100

1. The upper layer formed by semi-shrubs;
2. The second layer formed by small herbs (*Eremopyrum buonapartis, Carex pachystylis*, etc.).

Vegetation of salt deserts

In salt deserts the following plants are typical: *Halocnemun srobilaceum*, *Nitraria schoebery* and *Kalidum caspicum*. These communities are confined to soils rich in water soluble salts. Plants belonging to the group of halophytes are common here. Such species as *Nitraria schoebery* can grow normally in soils containing 7.8% Cl and 12.3% SO_4 (Korovin, 1961). Other plants can't exist here. Halophytes usually don't form closed communities. But in some cases one can meet dense patches of halophytes on solonchaks.

Vegetation of loess deserts

Loess deserts are located in foothill country. The highlands of Badkhyz and Karabil in Turkmenistan are typical habitats of this formation. This type of landscape is known in Turkmenistan as "adyr" and in Uzbekistan as "bair". Plant communities are formed here mainly by herbs (*Carex pachystilis, Poa bulbosa, Eremopirum triticeum, Aegilops juvenali, Psammogeton canescens, Astragalus* sp., etc.). *Ferula badrakema* and *Dorema badkhysi*, high ephemerals from *Umbeliferae* family form the first layer. They give a specific aspect to this landscape. *Artemisia turanica* and *A. badkhysi* also form several communities in this area.

Tugai vegetation

Vegetation of river valleys is known in Central Asia as "tugai". The areas occupied by tugai jungle have been declining from year to year because of severe human pressure and agricultural reclamation. Korovin (1961) distinguishes in tugai vegetation three formations: forests, shrubs, herbaceous vegetation. The most common species are the following: *Populus sp., Salix sp., Tamarix sp., Lycium turcomanicum, Halimodendron halodentron*, etc. More than 70 herbaceous plant species form here dense thickets (Figure 2.9).

Mountain vegetation

Mountain vegetation in Central Asia includes about 5,400 species (Korovin, 1961). Distribution of plant communities is strongly correlated with vertical zonality, specific climatic featu-

Figure 2.9. Tugai jungle in the Tedzhen river valley, Turkmenistan. Photo by A. Prishchepa.

res and the history of the flora. The mountains of the region are at present breeding grounds for formation of new plant species. The following vegetation types can be distinguished here:

- Mountain semi-savanna which is mainly met in the Pamiro-Altai and the West Tien Shan. The main plants are the following: *Alberta paleacea*, *Acanthophyllum panuculatum, Korovona tenuisecta, Agropyrum trichophorum, Ferula jaeschkeana, Bromus inermis, Eremostachis lemanniana*, etc.;
- Mountain steppe is mainly formed by cespitose xerophytes. The belt of this steppe is located to the south of 48° N. Local people call this landscape "chul'". The species of *Stipa* genus play an important role in formation of plant communities, in particular: *Stipa capillata, S. grandis, S. lessingiana*. Several other dominat plant species should be also mentioned: *Alium caesetum, Artemisia dracunsulus, Caragana aurantica, C. kirghizorum, Euphorbia alavatica, Cousinia sp., Polygonum hissaricum*, etc.;
- Mountain xerophytes. Plant communities of this formation are met at different elevations above sea level, up to sub-alpine zone. Plant cover is formed by the following species: *Acanthophyllum microcephallum, A. glandilosum, Convolvulus traganthoides, Schrenkia kultiassovi, Onobrichis echidna, Scorzonera acanthoclada*;
- Mountain deciduous forests. Their upper limit runs at the elevation of 2000 m above sea level. They are found in patches because of high human pressure. More than 100 tree and shrub species belonging to the following genera are registered here: *Acer, Betula, Lonicera, Malus, Pyrus, Cerasus, Armeniaca, Berberis, Cotoneaster*, *Padus, Eunimus, Fraxinus, Sobrus, Populus, Rubus*, etc.;
- Mountaim coniferous forests. They grow at elevations of 1800 – 1900 m above sea level. *Picea schrenkiana* is the dominant species in the Tien-Shan mountains. In the Kopetdad, Turkestan ridge and other mountain areas, the plants of *Juniperus* genus, totaling 11 species are common *(J. turcomanica, J. tienschanica, J. talassica* and other). All these species bear the local name of "archa";
- Shiblayk (thickets of shrubs and trees). *Pistacia vera* is a typical plant of this formation. Stands of *Pistacia vera* have been intensively exploited by man. This tree is confined to the lower sub-arid belt of mountains. Other edificators should be mentioned here: *Amygdalus communis, Punica granata, Rhus coriaria, Zizyphus coriaria, Paliurus spina christi, Ficus cariaca, Celtis caucasica, Cercis griffithi, Ephedra intermedia;*
- Mountain meadows, more than 100 species participate in forming these plant communities.

2.5. Livestock raising

The distribution of agricultural lands of Central Asia, by land use types, is given in Table 2.4.

Liverstock raising was from ancient time a dominant branch of the national economy (Figure 2.10). The natural fodder can be used all year round. Snow is usually not deep. Often it is melting under the warm winter sun. But because of low rangeland productivity, flocks should be driven from one place to other. One head of sheep needs 6 –10 ha of grazing land (Central Asia, 1968). A sketch of the rangeland management system is given in Figure 2.11.

Table 2.4. Agricultural land types of Central Asia, thousand hectares (Babaev & Kharin, 1994 b).

Land types	Kazakhstan	Kirghizstan	Uzbekistan	Turkmenistan	Tajikistan	Total
Arable lands	36000	537	4184	1233	822	42775
Irrigated lands	2000	404	4224	1331	713	8672
Haymaking lands	300	166	119	9	30	624
Grazing lands	180000	5223	23144	34527	3435	246339
Total	**218300**	**6340**	**31670**	**37610**	**5000**	**298410**

Rangelands of sandy deserts carry a great variety of forage. Spring is the best season of grazing. The carpet of ephemeral herbs covers the soil surface. Sheep are fed on *Carex physodes* and annuals. Summer is not favorable for grazing. But animals being fattened in spring don't suffer from heat in summer. The ration of sheep in this season consists of dry shoots of *Carex physodes*, twigs and seeds of *Calligonum* species and summer herbs. In autumn, sheep feed on twigs of *Carex physodes* and twigs and seeds of desert shrubs, the productivity of forage in sandy desert ranges from 60 to 200 kg/ha, depending on weather conditions (Nechaeva & Nikolaev, 1962).

Rangelands of gypsum deserts are divided in two groups: one with dominance of *Artemisia* species and other with dominance of *Salsola* species. Rangelands of the first group are suitable for sheep and camels and rangelands of the second group only for camels. Camels cover great distances feeding in spring on *Artemisia kemrudica*, *Salsola arbuscula* and ephemeral herbs, in winter they mainly consume the *Salsola* species. Rangelands of the second group are grazed by sheep only in years of plentiful precipitation when ephemeral herbs are abundant. Mean annual production of forage on these rangelands ranges from 60 to 200 kg/ha.

The main forage species of loamy deserts include *Artemisia herba-alba* and rarely *Haloxylon aphyllum*. In spring, sheep consume *Artemisia* species and annual herbs; in summer, *Artemisia* species and annuals from *Salsola* genus. Annual forage ranges here from 80 to 420 kg/ha.

Rangelands of loess foothill deserts are located at elevation of 700 – 1100 m above sea level. A dense cover of herbs, especially in wet years, is a characteristic feature of this land. Herbs often have secondary growth, and on the account of that, productivity can increase1.5 – 2.0 times. During 3 – 4 years per decade, haymaking is possible here. Productivity ranges here in the limits of 150-310 kg/year/ha.

Stock raising is also a common practice in Central Asian mountains. Vast grazing lands are located in the Tien Shan, the Pamiro-Alai and the Western Pamir at the elevation of 700 – 4000 m above the sea level (Central Asia, 1968). Some 50% of these lands are summer rangelands, 30% are autumn – winter rangelands and the rest are used in winter. Some of these areas are covered with forest or shrub vegetation, being registered officially as forest land. Yak and its hybrid with cow are common domestic animals in high mountain areas. Yaks don't need additional forage in winter and can live in the open air the whole year round. People also use the yak as a pack animal.

Stock raising in the desert always depends upon the weather conditions. Animals could perish from a long frost, snowfall or drought. These catastrophes were called in Central Asia "dzhut". But the nomadic economy always has a chance to survive. Some animals always survived and that gave impetus to rehabilitation of flocks. This rehabilitation didn't need a special capital investment. So, as a rule, the traditional system of nomadic stock raising was always profitable. Livestock number is given in Table 2.5

These data relates to 1993, without taking into consideration possible falsification of information (so called "pripiska").

Planned economies were introduced in Central Asian Republics during the Soviet period. The task of this system was to increase animal pro-

Figure 2.10. Grazing sheep in the vicinity of settlement Yaskhan, Karakum desert. Photo by A. Prishepa.

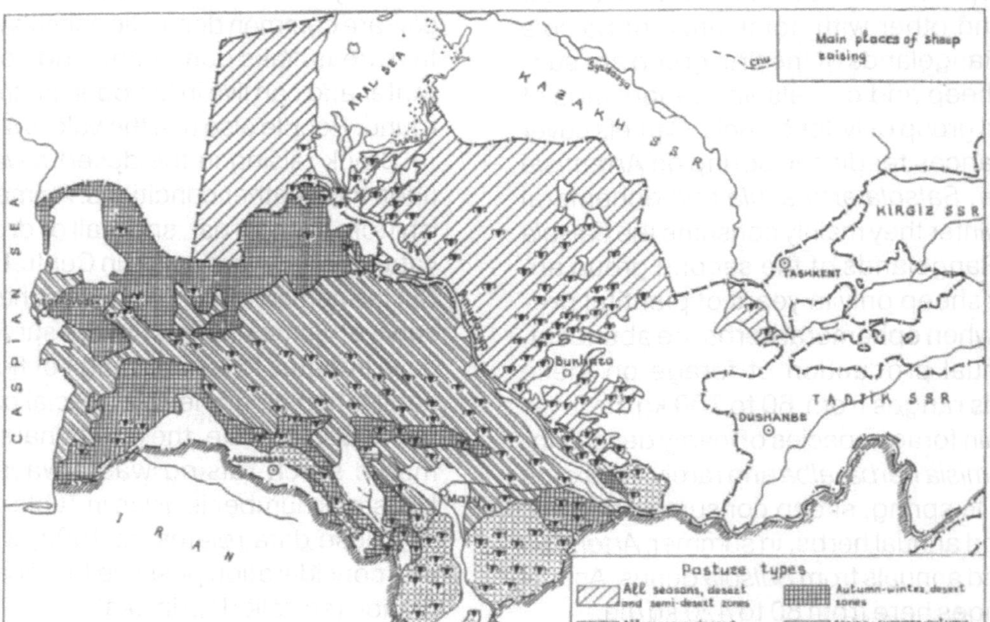

Figure 2.11. Grazing system in Central Asia (Central Asia, 1968).

duction by any way. It was supposed that this task could be solved by forage production on irrigated farm lands and by better rangeland management. The dynamics of livestock in two republics of Central Asia is given in Table 2.6. This table shows that the increase of livestock was not stable. In some years with cold winter or during droughts, the population of animals was reduced. For example, in 1979 – 1980, the winter was very cold in Central Asia and many hundreds of domestic animals died.

Meat and fat breeds of sheep were always raised in Central Asia. Local "Kurdyuk" (fatty tail) sheep have mean weights of 48 kg and produce 2 – 2.5 kg of wool. A new race of meat and fat sheep was bred in Central Asia during the Soviet period. The weight of this sheep totaled 65 kg, the wool clip ranged from 3.5 to 4.5 kg (Central Asia, 1970). Main products of livestock raising in Central Asia countries are indicated in Table 2.7.

Communist leaders of the USSR demanded an increase in the profitableness of livestock raising in the desert zone. That could be achieved by two ways: by increase of livestock number and by introduction of more profitable breeds of sheep. As known, karakul (Astrakhan) sheep races were bred before the Bol'shevik revolution in the Bukhara and the Khiva Khanates. So, during the Soviet period the number of karakul sheep had been increasing from year to year. That was the reason why the production of meat and wool in the deserts of Central Asia fell.

Table 2.5. Livestock number in Central Asian countries, thousand head (Babaev & Kharin, 1994c).

Country	Cattle	Including cows	Pigs	Sheep & goats	Horses	Camels	Poultry
Kazakhstan	9818.4	3326.8	3264.2	36222.7	1618.0	142.5	286.4
Kirghizstan	1214.3	507.4	444.8	10483.0	310.0	310.0	15206.7
Uzbekistan	4180.2	1644.7	742.6	8756.6	97.4	-	-
Turkmenistan	829.0	331.0	267.0	5481.0	19.0	87.0	6900.0
Tajikistan	1400.0	500.0	200.0	3200.0	-	-	8100.0
Total	17441.9	6309.9	4918.6	64143.3	2044.4	539.5	30493.1

Table 2.6. The dynamics of livestock (sheep + goats) in Turkmenistan and Uzbekistan during 1913 – 1980, mln. heads (National Economy of the USSR, 1972, 1991).

Republics	1913	1922	1924	1940	1965	1971	1980
Turkmenistan	4.6	1.9	2.4	2.6	4.0	4.0	4.5
Uzbekistan	4.3	3.3	4.1	5.8	8.0	7.9	9.0

Scientists of Turkmenistan bred new races of karakul sheep. According to Kurbanov (1984) the main ones are the following:

Arabi (black),
Shigari (grey),
Guligaz (rose-colored)
Kambar (brown),
Khalili (white),
Shaturi (the color of wool of a young camel).

Some 1.2 mln. skins of karakul were produced in Turkmenistan in 1980s. But according to one publication (Anonymous, 1985b) 2/3 of the karakul skins in Turkmenistan produced in 1984 were below international quality standards.

2. 6. Crop production

Irrigation agriculture

Gross production of agriculture in Central Asian countries is given in Table 2.8. Development of irrigation agriculture in Central Asia after 1917 was connected with ambitions of communist

Table 2.7. Main products of livestock raising in Central Asian countries (Babaev & Kharin, 1994c).

Country	Meat (in slaughtered weight), thousand tons	Milk, thousand tons	Eggs, mln.	Wool (in physical weight), thousand tons	Karakul, thousand pieces
Kazakhstan	1573.4	5562.9	4253.0	109.0	1833.7
Kirghizstan	241.4	1202.3	704.1	38.6	-
Uzbekistan	477.8	2929.3	2429.1	24.4	1483.0
Turkmenistan	104.0	435.0	328.0	16.0	574.0
Tajikistan	110.0	571.0	555.0	5.1	-
Total	**2506.6**	**10700.5**	**8269.2**	**193.1**	**3890.7**

leaders who demanded that cotton production be increased by any means. The Party's slogan was: "Give more white gold to the mother land!". Cotton was first of all a strategic raw material. Satisfaction of the needs of the Soviet people in textiles was the second, not the main task. The people of Central Asia plains dealing with production of cotton experienced the double pressure of high and local administrations demanding the growth of production of this crop. The dynamics of cotton production in two republics of Central Asia, where cotton growing is the main branch of agriculture, is given in Table 2.9. Table shows that by 1971 the cultivated area of cotton occupied more than half of the total cultivated area.

It is well known that cotton plants exhaust the soil. Chemical fertilizers have to be used constantly. Nitrous and potash fertilizers are mainly applied. Cotton plantations also need herbicides, defoliants and stimulators of plant growth. Cotton plants need much water. During a growing season cotton plantations should be irrigated 7 times. One hectare needs 7-8 thousand m^3 of irrigation water per season (Central Asia, 1968).

The problem of water was always very acute in the oases of Central Asia. Shortage of water was a serious obstacle to cultivation of cotton. But plenty of water was not good either. Turkmen proverb said: "Not soil gives the yield, but water". During ancient times this proverb was very true because of the shortage of water. But during the Soviet period this thesis played rather a negative role. Plentiful irrigation led to the secondary salinization. Very often Soviet "merabs" in collective farms observed the old tradition and gave much water to irrigated crops.

In Central Asian Republics, irrigation canals were usually constructed without any revetment, so some 40-60% of water was lost. That caused the uprising of saline ground water. Horizontal pipe drainage or vertical pump drainage were constructed to conduct this saline water from oases. But as a rule these systems were functioning very bad.

But we can mention one good example of construction of irrigation system in the former Soviet Union. The project of land irrigation in Golodnaya Steppe (lit. "Hunger Steppe") be-

Table 2.8. Gross production of agriculture in Central Asia, thosand MT (Babaev & G. Kharin, 1994c).

Country	Grain	Raw cotton	Potato	Vegetables	Maize for silos	Forage roots	Hay	Berries	Grapes
Kazakhstan	20,356.2	315.0	1,788.0	1,254.3	19,851.0	258.0	21,394.0	118.2	48.2
Kirghizstan	1,551.1	68.0	150.0	378.0	2,564.7	314.2	1,010.7	196.2	63.0
Uzbekistan	1,640.8	5,292.0	324.6	3,524.4.	-	-	-	550.0	416.0
Turkmenistan	462.0	1,457.0	35.0	726.0	1,027.0	44.0	793.0	45.0	152.0
Tajikistan	246.0	289.0	199.0	505.0	-	-	-	240.0	199.0
Total	**24,256.1**	**7,421.0**	**2,496.6**	**6,387.7**	**23,442.7**	**616.2**	**23,197.7**	**1,149.4**	**878.2**

Table 2.9. Dynamics of cotton production in Uzbekistan and in Turkmenistan (National Economy of the USSR, 1972, 1991).

Country	1913	1922	1924	1940	1965	1971	1980
Uzbekistan							
Sowing area of cotton, thousand ha	429	5 3	268	324	1550	1707	?
% from the total sowing area	19	3	19	30	46	50	?
Production of cotton, thousand tons	522	39	204	1368	3475	4511	5575
Turkmenistan							
Sowing area of cotton, thousand ha	69	25	40	150	257	401	?
% from the total area	22	14	17	36	50	60	?
Production of cotton, thousand tons	69	10	36	211	553	920	1192

longed to such a system. As known, reclamation of new lands in this region began before the Bol'shevik revolution (Palen, 1910a). But only after 1930 was a perfect irrigation system built. The Karshi main canal conducting water from the Amudarya to Karshi Steppe had good revetments and the efficiency of water delivery totaled 97%. According to Zonn, Nikolaev & Orlovsky (1981) the first development of the project on irrigation in Golodnaya Steppe included 400 thousand ha. Up to 1980 half of this area was reclaimed.

But drainage water from Karshi Steppe and from the Bukhara oasis was discharged into the Amudarya. The volume of this water totaled 0.57 km^3 and salinity was 5.45 g/l (Kolodin, 1992). So the water from the Amudarya became unsuitable for drinking. There was a plan to conduct this water (via the Amudarya by pipes) to the Main Turkmen collector drain.

Bureaucratism and formalism were characteristic features of the Soviet political system. That in full measure concerned the construction of irrigation systems. Dukhovnyi (1985) told about the formal approach in planning melioration works. Construction agencies received plans which included three separate tasks: leveling the irrigated fields, soil melioration, reconstruction of the irrigation network. These three tasks were considered as separate projects, and they were implemented separately, without any coordination. So, in several cases the final results were not achieved or they were not effective.

Without any doubt, the biggest irrigation structure in Central Asia was the Karakum Canal, named after V.I. Lenin. In 1992 it was renamed the "S. Niyazov's". The construction of the canal began in 1954. According to the project document, its total length would be 1340 km and the water discharge would reach 13.5 billion m^3 (Zakharchenko, 1984). Some 1 mln. ha of land suitable for irrigation were located in the canal zone. Additionally, it was planned to supply with canal water to some industrial enterprises and settlements. Fish-breeding was also possible in the canal and in water storing lakes.

A drainage network should be constructed to prevent the process of soil salinization. But this project was realized with great delay. For example, the Khauzkhan tract of new land had been irrigated since 1964, but the drainage network was constructed 10 years later. That resulted in intensification of the soil salinization process (Batyrov, 1992). About 15% of the underground drainage was not functioning in 1992. In such areas the loss of agricultural products totaled 60%.

The problem of drainage water discharge was not solved by construction of the Karakum Canal. A project of water discharge was developed only 30 years later. The question is of construction of the Main Turkmen Collector (MTC). This project was reviewed in detail by Kolodin (1992). According to the first outline, all three versions of MTC had to be discharged into the Caspian Sea. But the sanitary inspection raised objections against pollution of the Caspian Sea. So, the version No1 was approved. According to this proposal, drainage water would be discharged into the Kara Shor depression. The maximum discharge would total 16.65 km^3/year. MTC would go along the chain of Unguz solonchak where saline water would be cleaned by a hydrological method. Twenty nine natural depressions would be used as sedimentation ponds (Figure 2.12).

From the ecological point of view, this project was severely criticized. First of all, measures for prevention of water loss by filtration were not planned. More than 15% of water would be lost by filtration. Polluted water would mix with underground water. Water logging in the MTC zone would spread over the area of 1.5 mln. hectares, which represented 4.3% of the total area of grazing lands of Turkmenistan.

Drainage water from Tashauz oasis, from irrigated lands of Uzbekistan and from the Karakalpak republic has been discharged into the Sarykamysh lake. This lake, 26.6 km^3 in volume, was created by man on the account of drainage water discharged from irrigated lands.

Some time ago construction began of the Transkyzylkum Collector from Termez to the Aral Sea. This collector would transfer water from irrigated fields of Uzbekistan. It would also stimulate the regeneration of the Aral Sea.

Dryland agriculture

The areas of dryland agriculture in Central Asia are comparativly small. Patches of rainfed agriculture are mainly confined to piedmont plains. Right after soil thawing begins, cereal crops can be sown, and in the last days of April or in the beginning of May crops are ready for harvest. According to the moisture conditions, dryland agriculture is divided into three types:

1. Not guaranteed bogara in the areas with precipitation rate of 250 – 300 mm/year.
2. Bogara, half provided with moisture, precipitation ranges here from 350 to500 mm/year.
3. Guaranteed bogara in the areas where precipitation exceeds 500 mm/year.

The areas of dryland agriculture were especially enlarged at the beginning of the 20th century when many migrants came to Central Asia. These lands totaled 335 thousand ha. Even in 1943 bogara occupied 1.466 mln. ha in Uzbekistan (Central Asia, 1968). But soon after that, irrigation agriculture became the main branch of agriculture and bogara was abandoned.

Nowa days some 7.3 mln. ha, suitable for bogara cultivation are located in the Aral Sea basin. About 50% of this area belong to guaranteed bogara, but only a part of this land reserve is cultivated.

2.7. Pollution of environment

There are three main problems of atmospheric change in the region: increase of CO_2 concentration, freons in the upper layer of atmosphere and pollution of air with industrial gases and aerosols (Babaev & Kharin, 1994c). The sources of environmental pollution include: agriculture, chemical industry, production of fertilizers, production of building materials and others. Table 2.10 gives data on use of pesticides in agriculrure.

Turkmenistan and Tajikistan stay in the first place on application of chemicals in agriculture. The low level of agricultural technology and preference given to cultivation of cotton are the main reason for contamination of environment.

Table 2.11 gives information on environmental pollution by discharged sewage water.

The greatest amount of water without purification is discharged in Kazakhstan and Uzbe kistan. That is especially harmful for the people of Uzbekistan where the population density is the greatest in Central Asia.

Data on atmospheric pollution are given below (See Table 2.12).

Many towns and settlement of the region have high rate of atmospheric pollution. In 8 towns of Kazakhstan, including Alma Ata pollution rate exceeds the admissible level 3-14 times. In Kirghizstan two towns have the highest level of pollution, the pollutants being formaldehyde, carbon dioxide, benz(a)pyren and others. In Tajikistan Dushanbe heads the list of polluted towns. In 11 towns from 12 where observation were conducted, the content of dust and mean annual concentra-

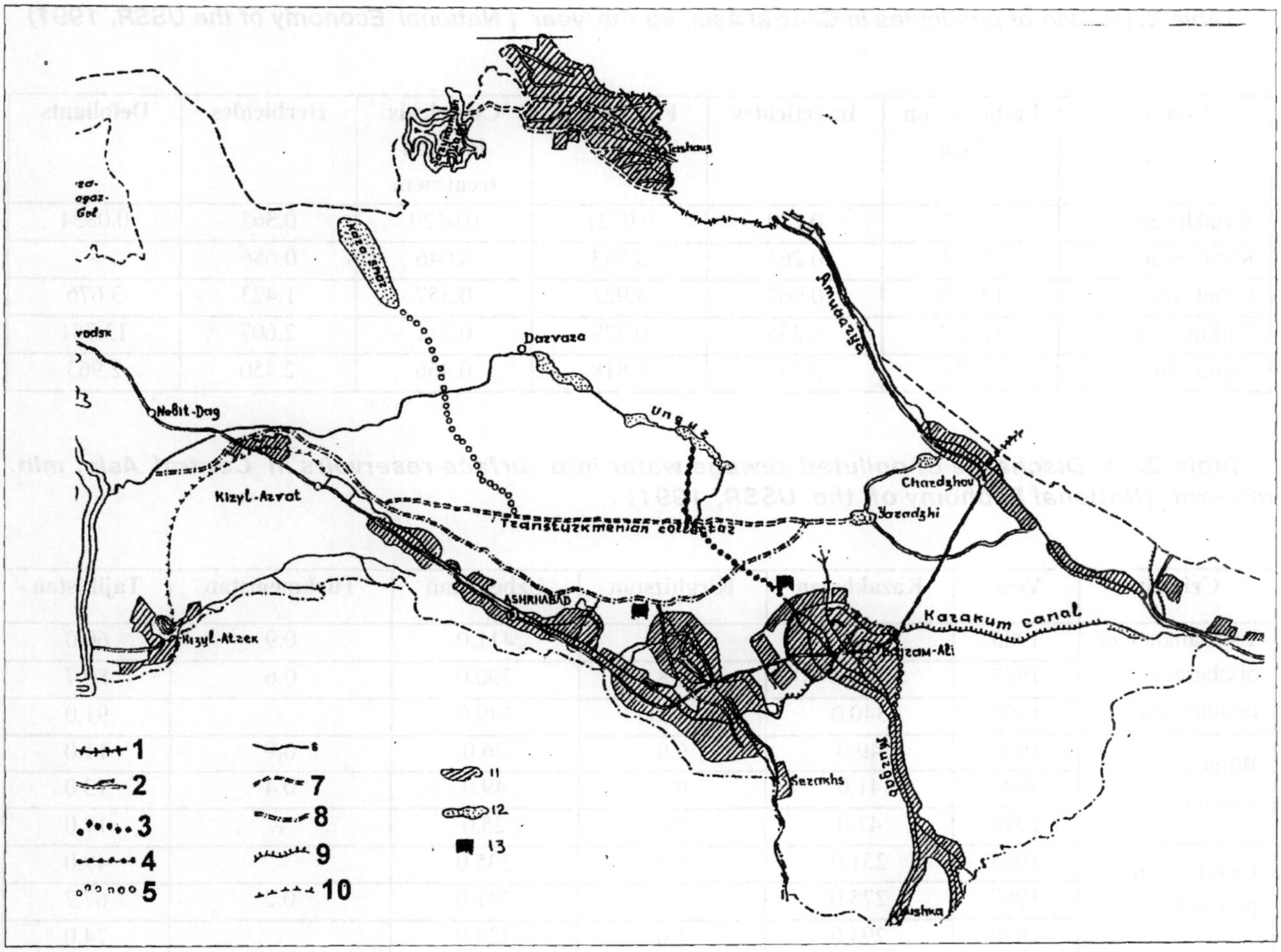

Figure 2.12. Drainage network in Turkmenistan.
1 – Inter-Republican collector, 2 – district collectors, 3 – acting Main Murgab collector, 4 – Main Murgab collector under construction, 5 – Turkmen collector under project, 6 – Trans-Turkmen joint collector (the first version), 7 – Trans-Turkmen joint collector (the second version), 8 – Trans-Turkmen joint collector (the third version), 9 – Karakum Kanal (the functioning version), 10 – Karkum Canal (under projection), 11 – irrigation lands, 12 – natural depression, 13 – desalinization plants, under project (Kolodin, 1992).

tion of pollutants exceeds the admissible level. In Turkmenistan, Chardzhou is the town with the highest level of pollution, exceeding the permissible level 10 times, Ashkhabad and Mary exceeds that 14 times. In Uzbekistan Fergana and Almalyk are the most polluted towns.

After reaching the upper atmosphere many pollutants after photochemically transformed into acids, salts and ozone. Fall-out of these pollutants may occur many hundreds kilometers from their sources in the form of acid rains. In 1989, the amount of sulphate fall-out in the region was 0.5 metric tons/km^2/year within the main part of the region. Mean annual fall-out of nitrogen in south-eastern regions of Kazakhstan, in Fergana valley and in Tajikistan totaled 1.0 metric ton/km^2/year (Babaev & Kharin, 1994c).

As known, continuous concentration of greenhouse gases in the atmosphere will result in global warming by 1.4 – 4.5 °C by 2000. It is supposed that temperature in the region will grow by 2 – 3 degree in winter and by 1.0 degree in summer by 2000. To bring down the risk of global warming, it is necessary to transfer from carbon-saturated fuels to those with high content of hydrogen (for example, natural gas). Alternative sources of energy will also give the same effect.

Table 2.10. Use of pesticides in Central Asia, kg /ha/year (National Economy of the USSR, 1991).

Country	Pesticides on whole	Insecticides	Fungicides	Chemicals for seed treatment	Herbicides	Defoliants
Kazakhstan	0.767	0.220	0.0721	0.0129	0.363	0.0854
Kirghizstan	2.028	0.263	0.953	0.046	0.686	-
Uzbekistan	12.360	0.967	5.922	0.357	1.423	3.676
Turkmenistan	17.443	0.235	0.529	0.211	2.607	13.781
Tajikistan	15.201	2.433	6.818	0.466	2.450	2.963

Table 2.11. Discharge of polluted sewage water into surface reservoirs in Central Asia, mln. m³/year (National Economy of the USSR, 1991) .

Criteria	Year	Kazakhstan	Kirghizstan	Uzbekistan	Turkmenistan	Tajikistan
Total amount of discharged polluted water	1985	280.0	12.0	411.0	0.9	60.0
	1987	316.0	5.8	390.0	0.6	80.7
	1990	340.0	14.0	349.0	-	91.0
Without purification	1985	49.0	6.0	76.0	0.9	11.0
	1987	41.0	0.5	49.0	0.4	13.0
	1990	47.0	9.0	25.0	-	17.0
Unsufficiently purified	1985	231.0	6.0	335.0	-	49.0
	1987	275.0	5.3	341.0	0.2	67.7
	1990	293.0	5.0	324.0	-	74.0

Environmental pollution and techogenic desertification severely increased in Kazakhstan during the last 10 years (Baitulin & Bekturova, 1997). Soil pollution is caused by atmospheric effluents from industrial centers (more than 4 mln. metric tons per year), from discharge of 6.0 cubic km of waste and from pesticides used in agriculture (up to 42 metric tons per ha on 26 thousand ha).The quantity of chemical pollution in soils around industrial centers doubled every ten years. Technogenic desertification was registered on an area of 181.3 thousand hectares, especially around industrial centers, and oil and gas pipelines. Space launching and military sites of the former USSR, occupying 6% of the Kazakhstan territory are also numerous foci of environmental pollution. The Caspian Sea has flooded 357 thousand ha of fertile land, range land and hay making land.

2.8. Summary and conclusion

The territory of Central Asia can be divided by physical characteristics into four sub-regions: steppe zone (not included in our study), semi-desert zone, desert zone, mountains. Each of these sub-regions has specific climatic features, soil and vegetation which form a zero level of desertification hazards. Human activities in development of desert lands and exploitation of natural resources disturb the environment and creates additional obstacles in desertification control.

Vulnerability of desert ecosystems to desertification is defined by several factors: aridity, mechanical soil composition (sand is the most vulnerable), soil salinity, stages of vegetation fluctuation, disturbances of plant cover and catastrophic changes in environment. Construction of the Karakum Canal and the drop of the Aral Sea

Table 2.12. Pollution of air and purification of harmful substance in Central Asia in 1990 (National Economy of the USSR, 1991).

Country	Total contents of pollutants, mln. MT	Contents of pollutants per capita, kg	Percentage of purification
Uzbekistan	5.0	255.9	69
Kirghizstan	1.2	274.8	84
Turkmenistan	0.9	254.0	44
Tajikistan	0.6	114	82

level are "good examples" of these catastrophic changes.

Inherent risk of desertification is especially high in the transition zones between physical landscapes (for example, foot-hill plains), between oases and desert land, in the Aral Sea bordering areas, and near the railways, motor ways, gas mains and industrial enterprises in the desert (Figure 2.13)

Desert vegetation, being a very sensitive indicator, suffers first of all from human pressure. As known, albedo is increased by destruction of plant cover, that stimulates the further development of desertification.

Environmental pollution and technogenic desertification increased during the last 10 years, especially in Kazakhstan. The quantity of chemically polluted soils around industrial centers doubled every ten years. Space launching sites of the former USSR are foci of environmental pollution.

Figure 2.13. Wind erosion near the gas-main in sandy desert.

CHAPTER 3

HISTORICAL INTERACTION OF DESERT AND SOCIETY

This is what the Lord said: The day when he goes down to the underworld, I will make the deep full of grief for him. I will keep back her streams and the great water will keep back her streams and the great water will be stopped: I will make Lebanon dark for him, and all the trees of the field will be feeble because of him.

Ezekiel 31:15

3.1. The first ecological catastrophe in Central Asia

Roughly 50, 000 years ago man had practically populated the whole territory of Central Asia. Archeologists named this period as a "Mustier Time". It was a primitive culture of hunters (Okladnikov, 1966).

An ancient culture of settled people , the culture of agriculturists was formed in the southwestern part of Central Asia in VI - V millennia B.C. According to Masson (1966a), this culture was named as "Dzheitun Culture". People cultivated barley and wheat. Stock raising was the second branch of economy, but this people got the main part of their meat by hunting. Dzheitun houses were built of clay blocks. The living room of single houses totaled 13 to 39 m^2.

Archeologists have also studied the relics of the Bronze Age (III-II millennia B. C.). One of them, the Namazga-Depe is located in Turkmenistan, southeast of Ashkhabad. Grains of several agricultural crops were found by excavations, in particular that of barley, wheat, rice, rye, chick pea and vine. Irrigation of fields was provided by runoff water and by water of small rivers running from the Kopetdag mountains. Draft animals, in particular camels, were used in agriculture. A primitive wooden plough was known to this people (Masson, 1971).

So called "Chust Culture" of farmers and settled cattle breeders was formed in the Fergana valley in the end of the IInd and beginning of the Ist millennium B.C. Archeologists have estimated by indirect indices that irrigation agriculture was developed by that time. The people cultivated barley, wheat and millet. They used bronze and stone tools in housekeeping. The Chust people bred cows, goats, sheep, horses, camels, donkeys and probably pigs. Flourishing agricultural oases of the Chust tribes fell into decay in the 7 – 8 centuries B.C. (Zadneprovsky, 1966).

As known, the flow of the Amudarya was changed many times in the past. The population of ancient Khorezm had many legends about the flow of the Amydarya through the Sarykamysh depression and Uzboi to the Caspian Sea. Herodotus said that Araks (the ancient name of the Amudarya) near its mouth was divided into 40 branches. One of these streams was discharged into the Caspian Sea, and others were lost in lagoons and marches. A new name of the river – Oksus became known after the campaigns of Alexander the Macedonian in Central Asia (Yusarov, 1995).

The ancient channels of Uzboi (Figure 3.1) have been conserved until the present days, as well as the ancient channels of the Amudarya, for example Daryalyk. Archeologists have found many vestiges of agriculture developed along these channels. These vestiges related to ancient and medieval Khorezm. Some historians believe that ancient culture was destroyed by wars.

The breaks of the Amudarya towards the Caspian Sea were registered during later epochs, in particular in the 14th century A.D. In 1878 a new channel of the Amudarya, 200 km in length was formed, and water flowing through it reached the Sarykamysh depression and filled it (Moser, 1885).

Modern scientific methodology can be used for reconstruction of ancient anthropogenic landscapes. This reconstruction was made by Volovik (1992). He used space photos and archeological data. According to this study, 28 to 40 thousand people lived on the territory of the ancient Khorezm from the 5th century B.C. to the 5th century A. D. They kept 365 to 400 thousand head of livestock and consumed 330 kg of grain per year per capita. Masson (1977) said that people of this region consumed not only grain products but also meat (300 – 400 g per day).

Decay of agriculture in the 6 – 5th century B.C. occurred on the territory of Central Asia. Irrigation systems of Kunyaurgench, Shakhsenem and other cultural oases were desolated (Mamedov, 1984).

We can suppose that this ecological catastrophe was caused by physical but not by anthropogenic factors, for example, by drought or by change of river channels. Chaikovsky (1884) adduced some arguments for interpretation of the ecological situation in the 6th or in the 5th century B.C. on the territory of ancient Turkestan. His hypothesis was based on two sources: Herodotus and the Bible. Herodotus, "the Father of History", told about a country of five rivers deepening into mountains. All these rivers formed a joint river system. According to the interpretation of Chaikovsky (1884) this story concerned Turkestan, which was a part of the Assirian Empire (Assur in the Bible).

The territory of ancient Turkestan consisted of two satrapies (14 and 16) populated by two tribes: Sarangenians and Felanianians. These two satrapies paid taxes to the central Assyrian government in sum of 900 talents of silver (about 15 metric tons in modern units of weight). Egypt being a poorer country paid only 700 talents.

Chaikovsky (1884) found, that the five rivers mentioned by Herodotus were the following: the Chu with the Issyk Kul lake (the main river) and its tributaries, the Amudarya, the Syrdarya, the Zeravshan and the Murgab (or probably the Sary Su). The Uzboi, being the continuation of the main river, discharged into the Caspian Sea (Figure 3.2).

As the Bible said (Ezekiel 30:1), the word of God was given to prophet Ezekiel in the 11th year (after the destruction of the Jerusalem temple, i.e., in 576 B.C.). Chaikovsky (1884) believed that Ezekiel described in an allegoric form the prosperity and the decay of Turkestan, a country which "...got strength from the waters "... (Ezekiel 31:4).

God's penalty befell this country (Ezekiel 31:15), (see the epigraph in Chapter 3 of this Book). Taking this country as an example the prophet warned Egypt for its excessive vanity and haughtiness (Ezekiel 32:15). The same misfortune could overtake Egypt if its conduct did not change.

Later, Kovalevsky (1924) found a report which in his opinion confirmed the hypothesis of A.Chaikovsky. He found this confirmation in the

Figure 3.1. Uzboi, the ancient river channal. Photo by A. Prishchepa.

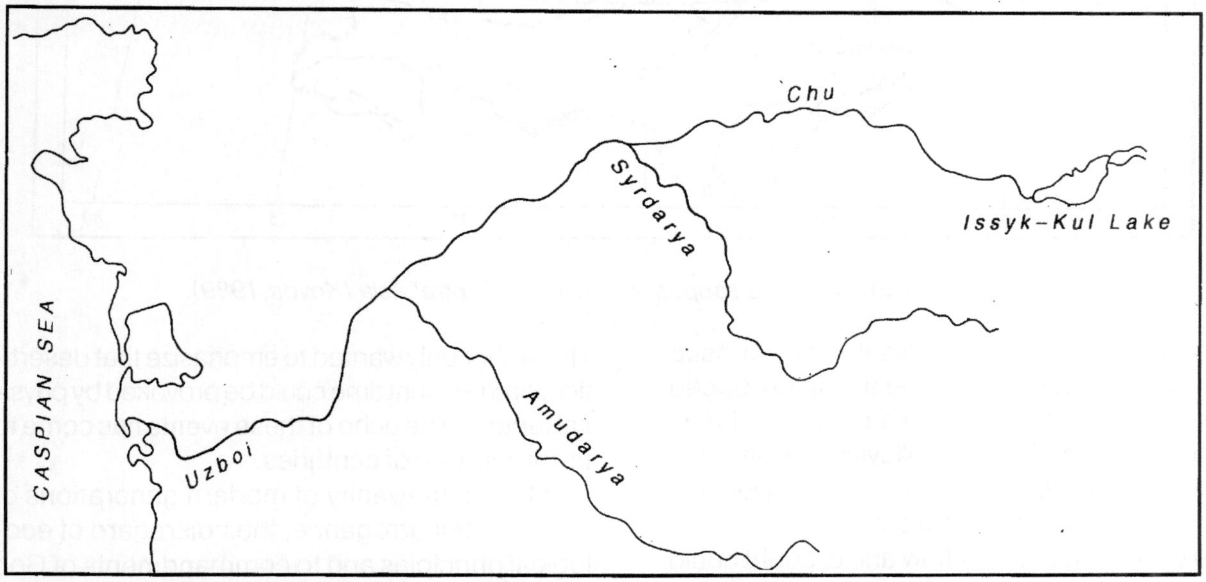

Figure 3.2. Reconstruction of the ancient river system in Central Asia by Chaikovski (1884), transferred to modern map by N. Kharin.

Veda, the sacred book of Hindus. He also believed that the water discharge of the Uzboi could total 2600 m^3/sec. But Kes' (1939), whose study concerned the problem of the Uzboi, refuted that. She said that the Uzboi couldn't have such a flow. In her opinion, the flow of this ancient river couldn't exceed 400 – 700 m^3/sec. At the same time it was possible, as she said, that the five rivers under consideration could form a joint river system.

Indirect confirmation of the Chaikovsky's hypothesis on the confluence of five rivers one could find in the map published by Kovda (1989). The map (Figure 3.3) shows the location of ancient river channels in Central Asia.

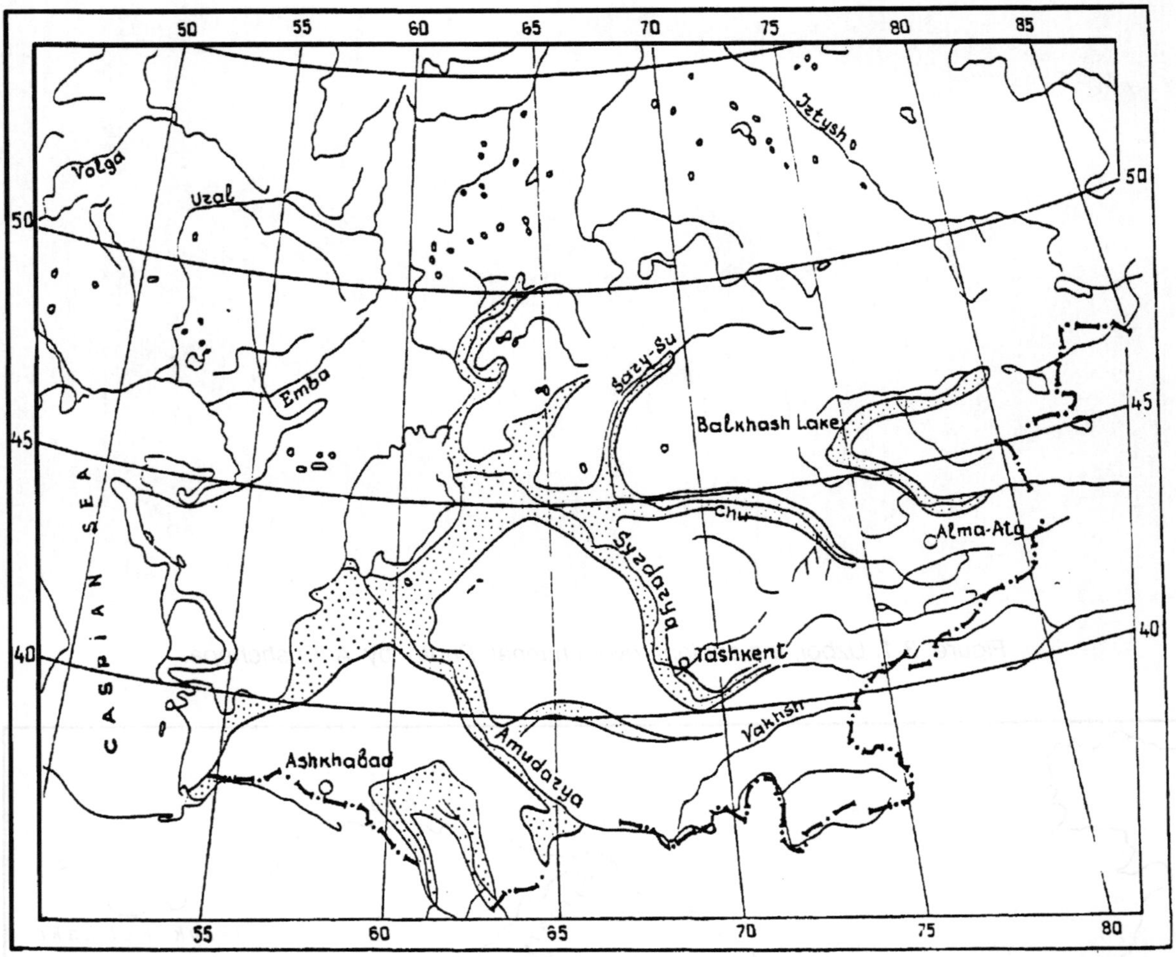

Figure 3.3. The areas of disappeared rivers of Central Asia (Kovda, 1989).

Korovin (1961) also published a map produced by Gerasimov (Figure 3.4). The map gives information about distribution of alluvial deposits during the latest alluvial phase. An ancient river system can be reconstructed from this map.

So the change of river flow and drought could be considered as causes of the first ecological catastrophe in Central Asia. More detailed study of this issue is beyond the scope of this book. The author only wanted to emphasize that desertification in ancient time could be provoked by physical factors. The echo of these events has come to us through the of centuries.

Maybe, the vanity of modern generations of people, their arrogance, their disregard of ecological principles and to commandments of God can be considered as ideological causes of desertification? In this case a warning to Egypt is pertaining to present days because it concerns all

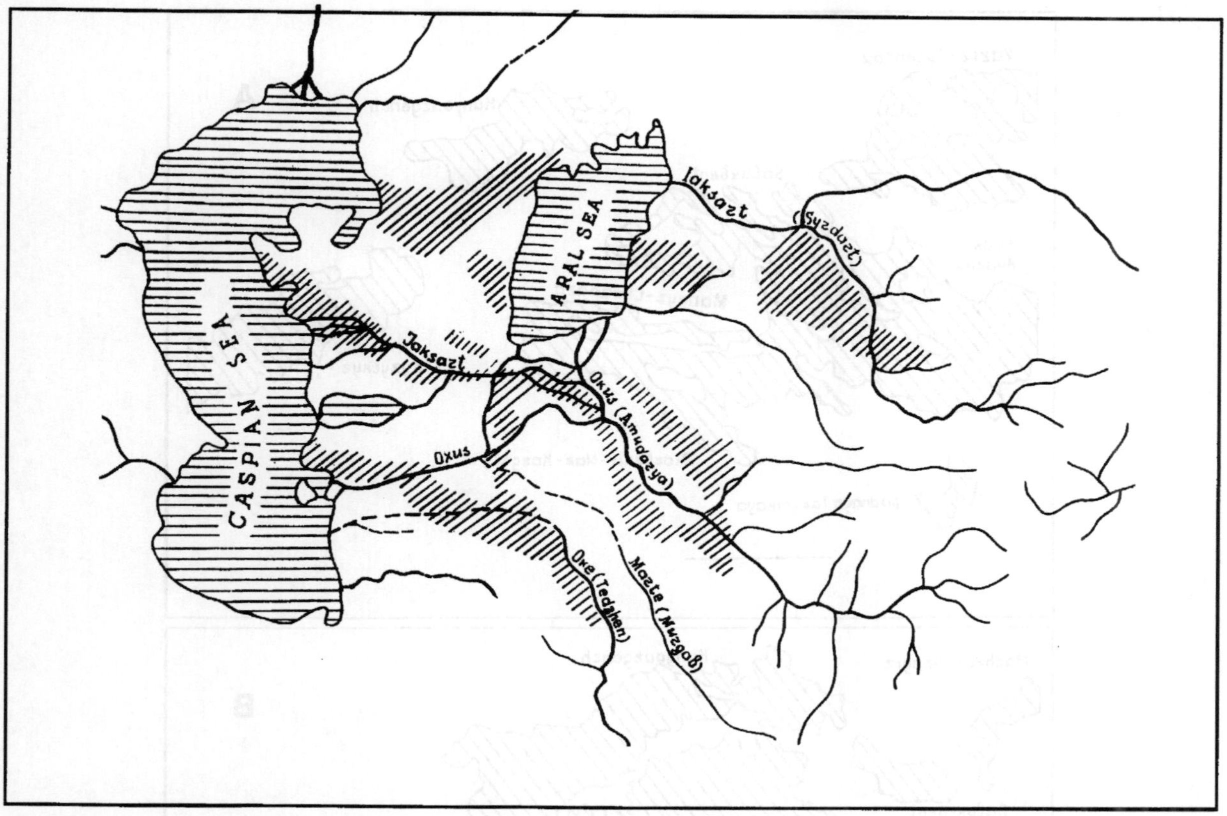

Figure 3.4. The areas of alluvial deposit in the later pluvial phase.

the people on the Earth: "When I make Egypt an unpeopled waste, cutting off from the land all the things; when I send punishment on all those living on it, then it will be clear to them that I am the Lord" (Ezekiel 32:15).

3.2. Land use and impact of war during the Middle Ages

Irrigation agriculture played an important role in the economy of Turkestan in the Middle Ages. Bartol'd (1900), in his monograph, gives the detailed characteristics of this problem. He describes the following periods in the history of Turkestan. The whole territory of Turkestan was conquered by Turks during 563 – 567 A. D. During 659 – 670 A.D. the region was a part of the Chinese Empire.

The Chineses were defeated by new conquerers, the Arabs. The State of the Sassanids was conquered by Turks at the end of the 10th century A. D. Khorezm became the capital of the vast empire which included the whole of Turkestan and Iran. Irrigation systems were often destroyed during bloody wars. So called "war desertification" provoked by conquerers devastated flourishing oases and was the cause of economic stagnation.

It should be emphasized here that normal functioning of the irrigation canals of Fergana, Khorezm and other oases was possible only by a strong centralized power. The heavy labour of great number of people, including slaves, was used for construction of canals and their cleaning. During the rules of the Sassanids, irrigation techniques were perfect developed. The irrigation system of the Murgab oasis made a great impression on Arabs. They even gave the name of Murgab to an irrigation canal in Basra.

Volovik (1992) made a historic reconstruction of anthropogenic landscapes in the northern part of modern Turkmenistan (Figure 3.5).

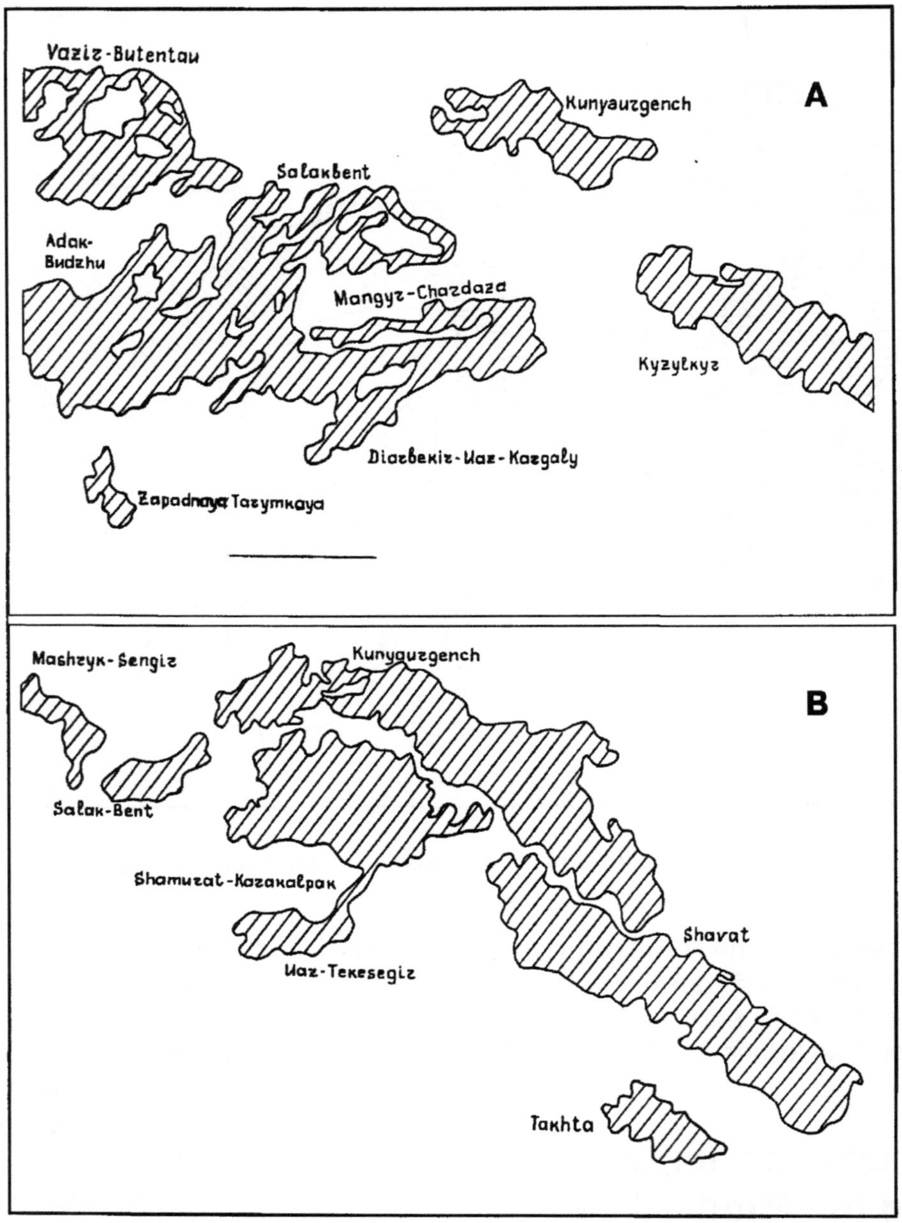

Figure 3.5. Location of irrigation lands in an experimenral area in North Turkmenistan. A – in the Middle Ages, 8 – 9th centuries, B – during the later Middle Ages, the 12 –17th centuries (Volovik, 1992).

This reconstruction embraced Urgench in the Middle Ages. Irrigated land occupied 363,309 ha during the 12 – 17th centuries A.D. The population was 78 – 90 thous and people, the mean population density being 46 persons per km^2. Livestock numbered about 1 mln. head. During the second period (the18 – 19th centuries), the irrigated area totaled 349, 872 ha, population increased to 110 thousand and livestock reached 1.2 mln. head. To our great regret, we don't have analogous information on other parts of the region.

The Mongolian conquest of Central Asia began in the second decade of the 13th century. Some historians believe that drought and summer snow falls gave impetus to movement of Mongolian hordes (Jenkins, 1974). According to his hypothesis the movement of Mongolian hordes could be provoked by the severe climatic changes, that occurred in 1175–1300 A.D. in Mongolia. The author explained that by the change of activity of glaciers in the northern hemisphere (See Section 2.2 of this Book). Fall of mean annual temperature was the reason for the dearth of forage in the Mongolian steppe. The number of wild animals during the same period also declined. As known, hunting was an additional source of food for Mongolian people. Without any doubt, the lack of forage could give an impetus for aggressive wars. But probably other reasons should be mentioned, first of all, the striving of Chinggis Khan for world suprimacy.

In 1211, advanced Mongolian detachments reached Semirechie (Bartol'd, 1914).The northern part of this area joined the Mongolian Empire. The State of Khorezm Shakh became a joint frontier with Mongolia. At the same time Mongolian hordes moved to China. That somewhat disracted them from Central Asia.Peking was captured in 1215. This event glorified Chenggis Khan.

In autumn 1219, Chenggis Khan, with numerous soldiers invaded Turkestan. By estimation of some historians the Mongolian army totaled 600 to 700 thousand of warriors. Bartol'd (1900) gives more limited figures. In his opinion the Mongolian army had 150 to 200 thousand warriors. The army of Khorezm Shakh was more numerous but because of feudal discord several detachments fought separately. That finally decided the fate of Khorezm Shakh Mukhammed and his empire. Mukhammed himself, pursued by Mongols, died on a small island of the Caspian Sea. *Sic transit gloria mundi...*

Chenggis Khan captured Bukhara in1220, then Samarkand was also occupied. Some 500 thousand people resided in Samarkand. An artful irrigation system was functioning in Samarkand. It was operated under supervision of fire – worshipers who for their labor were released from paying the poll-tax. Water to the town was conducted by a special canal Dzhakerdiza. The canal was also called "Leaden Channel" because it was faced by lead. Mongolian invaders destroyed the Leaden Channel and the dam. Subsequently the walls of the citadel were also destroyed.

By the way, the rulers of Khorezm used the same method to flood the narrow cultivated strip along the Amudarya when Mongols tried to capture the town. So in the both cases war desertification did harm to the environment and to the people.

Then Urgench was captured after bloody fighting and a long siege. Mongols destroyed the dams on the Amudarya, the river changed its channel and began to flow to the Sarykamysh depression. The town of Merv (now Mary) was one of the ancient hearths of culture in Central Asia. It occupied 2 square verst (about 4.8 km^2). The Mongols captured the town in 1221, destroyed the houses and murdered all the people. The number of killed people, according to different sources, totaled 700,000 to 1,300,000.

A flourishing town, Meshedi-Misriani was located in southwestern Turkmenistan before the Mongolian conquest. The town occupied an area of 130 ha (V.V. Bartol'd, 1900). It was surrounded

Figure 3.6. The ruins of Meshedi-Mesriani. Photo by A. Prishchepa.

by a high wall and a deep moat. Water for irrigation was conducted from the Atrek by a long canal, 60 km in length, 2 sagene (4.3 m) in width. Mongolian invaders destroyed the town, which was never restored again. The ruins of Meshedi-Misriani have been conserved until the present time (Figure 3.6).

During the Soviet period, vandals of the 20th century – gunners of the Red Army tried to complete the destruction initiated by Mongols. They used minarets as a goal and shelled the ruins. One of the shells made a hole in the tower of a minaret but the minaret itself was not destroyed!...The secret of the cement moisture used by ancient bricklayers has not yet been discovered.

Without any doubt, Mongolians were barbarians and cruel conquerors. But some reasonable regulations of their rule over subdued nations were remarkable and could attract our attention. Chenggis Khan died in 1227. His successor Ugedei became Khan in 1229. Kurultai of Mongolian tribes held in 1239 introduced the following taxes to be paid by subdued nations: 10% of yield for farmers and one head of livestock from hundred of animals for nomads. It would be nice if modern rulers of Central Asian countries followed these Mongolian regulations...

The dams of Merv were destroyed several times in the latest epoch. A cruel tyrant, Nadir Shakh came to power in Iran in the first half of the 18th century. He made several bloody raids to Central Asia. The Khiva Khanate was conquered by him in 1740. During this war irrigation canals and dams were destroyed in Merv and in other oases. Then he himself ordered the repair of all irrigation systems to provide a high tribute (Ivanov, 1938).

Historical sources give the following characteristics of the Merv oasis. The climate was dry, but the air contained harmful exhalations originated from water reservoirs located around the town. Many dangerous diseases caused by these exhalations were common in Merv. The land was fertile. In the first year of reclamation the yield of agricultural crops was 100 greater than quantity of seeds used, in the second year, the ratio of seed to crop yield totaled 1:50, in the third year 1:10. Cotton, rice, sesame and vegetables were the main crops (Ivanov, 1938).

In those times, the oasis of Merv was a center of developed agriculture. According to Mez (1922), the water management department of Merv had 10,000 employees dealing with distribution of water. A water measuring device was installed at a distance of 1 pharsaph from Merv. It was a board with a longitudinal cut in which a barley grain moved depending on the water level. If this indicator was on "60 grains" the people were very happy because in this case the quota of irrigation water was high and the yield of agricultural crops would be abundant. A level of "2 grains" was catastrophic because in this case people suffered from hunger.

The situation in sub-regions in the 18th century was unstable because of bloody wars and war desertification. Dwellers of Samarkand left the town for 7 years. Nearly all irrigated lands were neglected in the Khiva Khanate, and only 40 families lived in the town of Khiva. But in the beginning of the 19th century the situation was again stabilized. The Kokand Khans established in Fergana a rich state with developed irrigation system. This state existed till the Russian conquest.

3.3. Land reclamation, ecology and social life in Russian Turkestan

The 19th century in history of Central Asia was characterized by a relative stability in political life. Wasteful raids of conquerors, destruction of irrigation canals and mass massacre of people became a thing of the past. Local conflicts, like that between Turkmen tribes and the Khiva Khanate arose from time to time, but they couldn't disturb the whole stability in Turkestan.

The ecosystem of the Khiva oasis was less stable from the ecological point of view. Floods of the Amudarya occured here from time to time and the local population suffered from the change of river channels. But soon or later the dams were reconstructed and life in the oasis continued. But we must not forget that political systems in the

Khanates reflected the customs of true Oriental despots. Thousands of slaves worked on cotton plantations. Slave trade was a common business everywhere in the region. The first information about physical environment and customs of the local nations was published by travelers (Grigor'ev, 1869; Ujfalvy, 1880; Vambery, 1865).

Russian conquest of Central Asia

Conquest of Turkestan by Russia was without any doubt a great historical event. Historians, economists and the military assessed this event differently. Moser (1885) uttered his opinion based on a geographical approach. Frontiers of Russia in the south, from the Ural to the Don, were located in vast steppes where physical barriers were absent. The movement of such a great nation southeast was a natural movement up to the mountains of Afghanistan. That was the movement of a nation which wanted to make itself secure against raids of wild hordes of nomads. During the first period of Russian conquest, military expenditures were not justified. During 1868 – 1879 these military expenditu res for subjugation of Turkestan were as follows (in roubles):

Administrative expenditures	37,723,140
Military expenditures	75,108,019
Total	109,831,159
Return	43,015,219
Deficiency	66,815,940

From a military point of view the conquest of Turkestan was not a difficult task. Usually, the crowds of poorly armed natives scattered at the approach of Russian troops. But the main difficulty consisted in getting over vast areas of steppe and desert devoid of roads and water sources. Rickmers (1913) said: "It was the stroke of a giant on a rotten tree..."

According to other hypothesis uttered by Terent'ev (1906), expansion of Russia in Turkestan was caused first of all by political reasons. Russia was defeated in Crimean War (1853 – 1855) and the Russian tsar decided to take revenge in Central Asia. So the strike was made against the prestige of England in this region.

An economic point of view was also mentioned. According to Rickmers (1913), Russia needed raw materials and new markets. The War of Secession in North America (1861–1865) caused a decrease in cotton supplies, and the textile industry of Russia fell into decay. The conquest of Turkestan gave a new impetus for economic development of Russia.

It is very interesting to mention the main steps in the conquest of Turkestan. The Russian tsar, Nikolai the First, ordered in 1839 to build forts near the Amudarya. That would strengthen the position of Russia in Turkestan. The fortress at Vernoe was built in 1854 (now the city of Alma Ata or Almaty). Tashkent was captured in 1865. A treaty was signed between Russia and the Bukhara Khanate in 1868. Russian subjects could cross the territory of Bukhara without any obstacles. Free trade was allowed for Russian merchants, the tax should not exceed 2.5% of the total cost of goods (Moser, 1885). The Kokand Khanate was occupied by Russian troops in 1876. It received the status of "Fergana region".

In 1869 Russians opened a trading station on the coast of the Caspian Sea. Later the town of Krasnovodsk appeared in this place. A new name, "Turkmenbashi", was given to this town in 1992.

Then Russian troops began their movement to the east. In this movement the construction of the Transcaspian Railway played an important role. Rickmers (1913) said: "Siberia was slowly colonized, the Caucasus vanquished after bloody wars with the natives, but Turkestan was subjugated with comparative ease. The Transcaspian Railway was a piece of work almost American in the smartness of its execution, apart from the fact that Russian life and methods show many striking analogies with the ways of Uncle Sam".

Some Turkmen tribes, seeing an inevitable threat to their freedom, surrendered, as a matter of discretion. Any resistance was severely suppressed. A massacre made by Russian troops in the battle for the fortress Geok-Tepe testified to that.

The Politburo of the Communist Party in the Soviet period tried to falsify the history of the campaign for ideological reasons. And scientists obedient to the Party's bosses developed a legend about the voluntary entry of subjugated nations into the Russian Empire. For example, Gapurov, Roslyakov & Annanepesov (1983) said that 9/10 of territory of Turkmenia had joined Russia voluntarily. It was claimed that entry satisfied "age-long aspiration of Turkmen people to get Russian citizenship".

Turkmen "court poet" R. Seidov described the friendship between Russian and Turkmen people in the following way (our translation from Russian):

Our teacher and brother!
Thou the first embraced Turkmen,
We being the brethren
Pressed you to our hearts.
And you are doing the same
In response!...

Gapurov (1982), the former First Secretary of the Communist Party of Turkmenistan, cited these lines addressed to the Russians. But this myth was discredited. During the "perestroika" period, a special memorial stone was laid in Ashkhabad. A monument to mark the Russian-Turkmen friendship should be set up in this place. But the stone was secretly removed in the night after Turkmen independence was proclaimed...

But let's continue our story. The territory of Turkestan, as an administrative unit, was set up in the first period of Russian rule over Central Asia. The governor – genera l was the head of the administration, Tashkent being the capital. Firstly, the territory included two regions: Syrdarya and Semirechie. The Fergana region was joined in 1876 and the Samarkand region in 1887. The Transcaspian region, which included the territory populated mainly by Turkmen tribes, was brought into the administrative territory in 1899. Before that, this region was governed separately. So by the end of the 19th century, Central Asia, under the name of the Turkestan territory became an integral part of the Russian Empire.

The Russian administration conserved the existing local system of legal proceeding based on Islams's "Shariat" and "Adat". The Moslem religion, as a basis of the spiritual life of natives, was allowed to continue. And what is more, any missionary activities of the Russian orthodox clergy and the christening of Moslems were forbidden. Selling vodka to Moslems was also forbidden.

The wisdom of the national policy of "Ak Padishakh" was obvious. Ethnic clashes in Turkestan during the pre-revolutionary period were unknown in Central Asia. Strangers (not only Russians!) could travel everywhere without fear for their life and belongings (Rickmers, 1913). A strong centralized power and Russian troops were the guarantee of stability in the sub-region. The time of stability and prosperity lasted about 40 years, right up to the Bol'shevik revolution in 1917.

What was the economic situation in Turkestan during this period? Davletov (1977), a Soviet historian, found that the following measures taken by the Russian administration stimulated the development of Turkmen nation:

- Establishment of centralized admini tration,
- Abrogation of slavery,
- Confiscation of land from some feudal lords who possessed vast areas of land,
- Elimination of inter-nation feuds,
- Construction of irrigation canals,
- Construction of railroads,
- Creation of loan-banks.

Crop production

Irrigation agriculture has been carried out in Central Asia for centuries. According to encyclopaedia (Anonymous, 1902b), irrigated land in Turkestan totaled 1.5% of the land area in 1890, that made 2, 525, 000 ha (without the territories of Khiva and Bukhara). Dryland agriculture occupied 545, 000 ha. Count Palen (1910b), who conducted the inspection in Turkestan, gave more precise figures based on statistical reports of local administration. Irrigated land occupied the following area (given in dessiatina and recalculated to ha):

Semirechie region	763,193
Samarkand region	480,016
Syrdarya region	70,313
Fergana region	840,000
Transcaspian region	147,000
Total	**2,442,881**

All land in the Turkestan territory was declared a State property. The local people could use the land on the basis of existing regulations. The Russian administration stimilated the settlement of nomads. For example, the Kirghizes realized that numerous dzhuts could lead to their impoverishment. So the administration helped them to plow virgin land. This land belonging to settled Kirghizes covered in 1900 an area of 100,000 dessiatina in the Syrdarya region and 130,000 dessiatina in the Semirechie region (Anonymous, 1902b).

But the Russian administration did not arrive at this policy at once. According to Terent'ev, (1906) general P. Sukhtelen – by the rule of the Nikolai the First – didn't permit the Kirghizes to settle in the Orenburg region. He said that nomads were incapable of agriculture. Their task was to provide Russia with meat, wool and leather.

As Davletov (1977) stated, Turkmen tribes before the Russian conquest stayed on the patriarchal feudal stage of development. The land tenure system was a typical feature of this society. Land was a common property of local communities. The Tsarist government didn't change this system. At the same time, a new administration stimulated the development of agriculture by constucting new canals, karez, setting up loan-banks, etc.

Irrigation water in oases was distributed by aryk aksakals. They were responsible for distribution of water from main canals to small aryks. Merabs controlled the distribution of water from aryks to irrigated fields. The Russian administration transformed aryk aksakals to state officials. They were appointed by the administration and paid from the public coffers. Merabs were elected by local communities. Irrigation water was lifted from aryks to fields by primitive tools: chigir' and nova which were operated by horses, camels or donkeys. These tools have been existed from time imme-moria.

As Palen (1911) said, a fallow system was common in the agriculture of Turkestan. Arable fields were divided into 2,3, or more sowing areas which were annually sown in turn. In the Merv uyezd of the Transcaspian region irrigation fields sown by barley and wheat were irrigated 2 – 3 times, lucerne was irrigated 4 – 8 times. In Tedzhen uyezd wheat and barley were irrigated 4 times, in Krasnovodsk uyezd – 5 times. Ketmen' and omach were the main agricultural tools. Ketmen' was a kind of spade. This tool has been used in Central Asia until present days. Omach was a primitive plough with an iron ploughshare. At the present time, it is not used.

The people of Central Asia developed several original technique for tillage and constructing aryks. Let's consider only one example. Before constructing an aryk on a slope the levelling should be conducted. It was done by the help of heavily loaded donkeys. A man went in front of the donkey, a little lower on a slope. He didn't allow the donkey to go down and the obedient animal was moving exactly along a contour-line. The second man followed them and marked the line of the future canal by stalks of reed.

Palen (1910a) said that small land holdings dominated the agriculture of Central Asia (Table 3.1). This table demonstrates the history of agriculture in Central Asia. The most ancient tracts of irrigated land were located in river valleys and on piedmont plains. Because of population growth, these holdings were divided into small patches of land. In some districts there were diminutive holdings, of less than 1 dessiatina in size. Production of grain in Turkestan in the end of the 19th century is given in Table 3.2.

We have calculated the production of grain per capita per year, without oats (Table 3.2). According to FAO (Provisional methodology.., 1981) consumption of grain in a normal diet should be 224 kg per capita per year. From this point of view, the production of grain was insufficient in two regions: Samarkand and the Transcaspian. But the production of grain in the Turkestan territory as a whole was sufficient. In 1993 consumption of grain products in Central Asia per capita was lower. See section 8.3 of this monograph. Cotton was the most profitable crop in Central Asia. So the areas of cultivation of this crop increased from year to year during the period of development of capitalism in Russia (Table 3.3).

According to an encyclopaedia (Anonymous, 1902b), 339.5 thousand ha were in cotton in Turkestan in 1990, including the Khiva and the Bukhara khanates. Production of cotton to-

Table 3.1. Distribution of land holdings on size in three uyezds of Samarkand region, in percentage (Palen, 1910a).

Land types	< 1 dessiatina	1–3 dessiatina	3–5 dessiatina	5–10 dessiatina	>10 dessiatina	Total, %
River valleys	36.26	33.11	15.05	10.62	4.96	100
Steppe	3.79	15.45	17.94	35.57	27.25	100
Piedmont plain	18.09	30.80	19.75	18.27	13.09	100
Mountains	2.21	22.15	30.39	37.57	7.68	100

Table 3.2. Production of grain in Turkestan, metric tons (Anonymous, 1902 b).

Regions	Spring rye	Winter wheat	Spring wheat	Barley	Oats	Millet	Rice	Total	Per capita, kg/year
Semirechie	6,716	3,276	18,116	45,618	57,821	52,416	18,182	202,145	223.9
Samarkand	-	1,114	73,710	65,520	-	-	57,330	197,701	295.9
Syrdarya	-	87,142	79,034	66,748	-	62,224	89,762	384,930	260.1
Fergana	-	11,867	48,892	1,556	-	13,189	60,606	136,110	157.3
Transca-spian	-	6,880	20,484	9,418	-	-	-	36,782	123.0
Total	**6,716**	**110,279**	**240,236**	**188,860**	**57,821**	**127,849**	**225,880**	**957,668**	**243.1**

Table 3.3. Increase of areas sown by cotton in Turkestan, hectares (Davletov, 1977).

Years	Samarkand	Syrdarya	Fergana	Transcaspian	Total
1890	18909	25615	55744	981	110125
1895	16593	15374	119574	3815	153355
1900	24879	16484	203095	10900	255363
1905	20423	16690	181799	11990	230902
1910	27494	38886	257121	30520	354021
1915	60575	80715	366812	62566	570668

Table 3.4. Production of cotton in Turkestan, metric tons (Yuferev, 1925).

Years	Samarkand	Syrdarya	Fergana	Transcaspian	Total
1907	2,850	5,127	85,176	438	98,591
1910	10,712	15,240	116,020	11,794	153,766
1913	15,610	19,771	124,652	14,234	174,267
1917	13,530	11,220	61,687	6,650	93,087

taled 75 670 metric tons, the mean yield of cotton per hectare was 0.22 metric ton.

Production of row cotton in four regions of the Russian Turkestan before the Bol'shevik Revolution is given in Table 3.4. During World War I and during the civil war, production of cotton fell severely.

It should be taken into consideration that in 1913, in the Bukhara and the Khiva khanates, about 100 thousend hectares of land were growing cotton, and production of cotton totaled 13,270 metric tons. This information was not included in Table 3.4.

We don't have detailed information on salinization of irrigated lands in Turkestan. But as known, ancient irrigation canals were constructed without lining and water loss in irrigation systems could total 60% on account of evaporation and leaking. People trying to get more production applied much water to irrigated fields. A Turkmen proverb said:"Not earth but water bears the yield". As a result the level of ground water rose and salts also moved from deep soil horizons nearer to the earth surface.

Dekkhane developed a local technique to control the process of soil salinization. Salt crust on the earth surface was the first indicator of soil salinization. This crust was removed from soil every year. Then, soil was covered with so called "weathered earth" taken from old duvals, depe and the ruins of saman houses. After that, fields were leveled and manured. This heavy work was conducted without use of any machinery. Cleaning irrigation canals from silt was also a heavy labour. Each dekkhanin spent for this cleaning 100–120 days annually (Khodzhamuradov, 1984).

The Russian administration tried to modify the existing system of land use. Russian officials conducted a land inventory and collected statistical information. But they couldn't change a traditional system of rangeland exploitation. Palen (1910a) said that "Temporary Code on Government of the Transcaspian Region" recommended to conserve the traditional system of moving from place to place.

Natives said, under questioning, that non-irrigated land was distributed among tribes "in ancient time". They didn't depend on documents nor any written papers for making decision. All was based on existing traditions and associated rules.

Traditional system of animal husbandry

A traditional system of distribution of grazing land was the following. Desert land was distributed among native tribes according to dug wells. The wells were the collective property of a tribe or kin group. But neither kin nor single persons had the right of property on land. Owners of wells had a preference for utilization of rangeland around the radius of 15 verst. They couldn't remove otara of other people from this territory, but they had a right not to give water to stranger's animals.

Desert stock – raisers roamed during the pre-revolutionary period for long distances. They began their transhumance in spring, moving to the north and following "the green forage". In autumn the flocks came back (Figure 3.7). The length of these transhumances there and back could totals 2,000 km (Nikolaev, 1972).

There was a special system of livestock grazing around desert wells. Each chaban tried to fatten all animals at an equal rate. His skill was based on the following approach (Nechaeva, 1953). As known, by movement of otara the stronger sheep went in advance, and the weaker sheep followed them. If chaban did not regulate the movement of animals, the stronger sheep would be well-fed, the weaker sheep would be skinny. So the main task of chaban was to create equal conditions for fattening all animals. Chaban turned otara round and sheep changed their places. As a general rule, sheep turned from the right to the left, and the whole otara moved counter clockwise.

Livestock number in Turkestan is indicated in Table 3.5. Development of livestock raising in the desert depends upon the weather conditions. Livestock needed grass all year round. The number of livestock herded by nomads was never stable. Desert dwellers always maintained as many animals as it was possible. Livestock were for nomads the main riches, their gold and currency. Dearth of fodder (dzhut) was a real scourge for them. During very severe years, win-

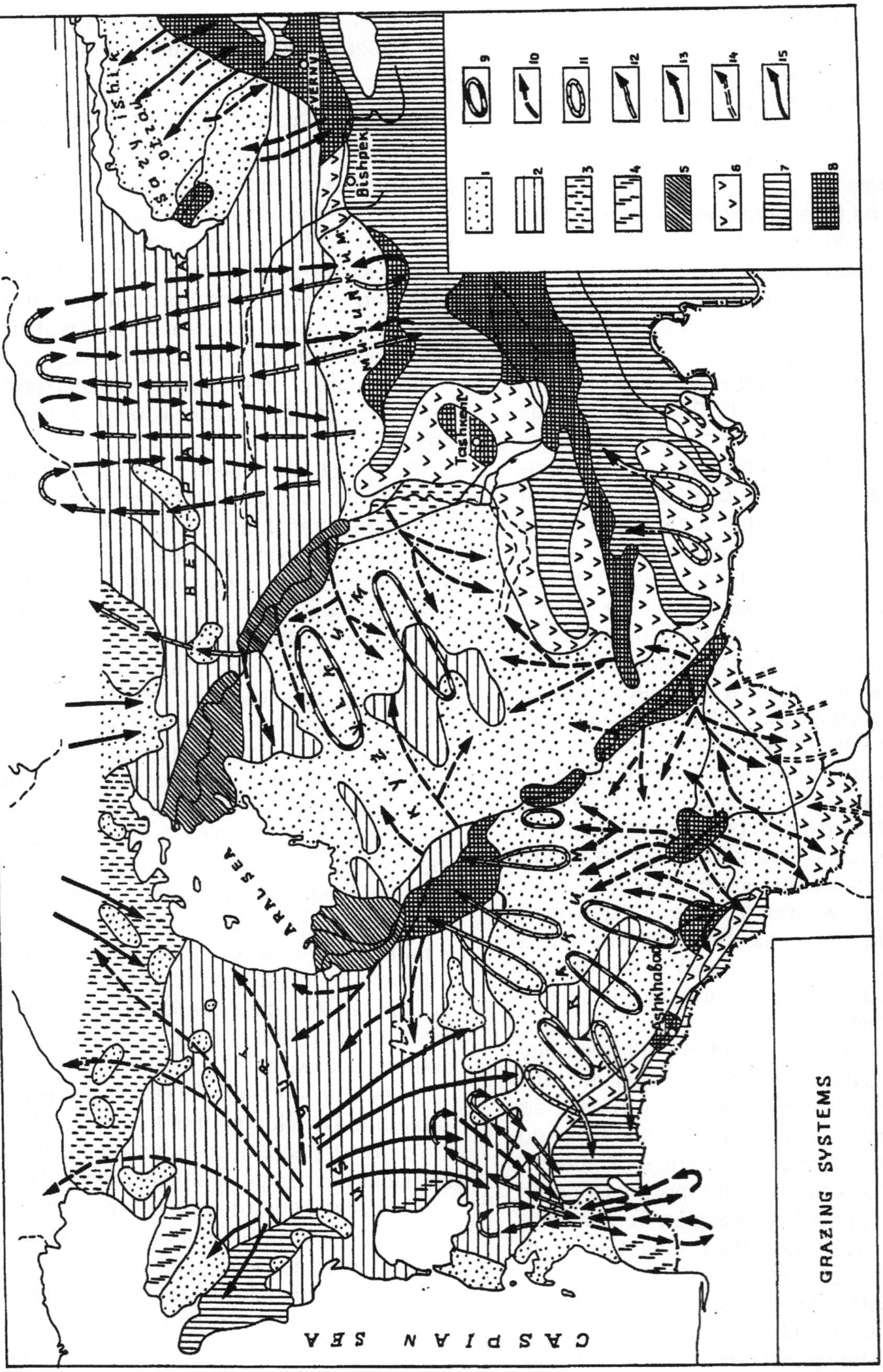

Figure 3.7. Transhumance system by livestok raising in Russian Turkestan.
Types of deserts: 1 –sandy, 2 – gypsum, 4 – solonchak, 5 – river valleys, 6 – foot-hill plains, 7 – mountains, 8 – oases. Seasonal movements of flocks: 9 – all round year, 10 – spring – summer – autumn, 11 – autumn – winter – spring, 12 –spring – summer, 13 – autumn – winter, 14 – spring, 15 – summer (Nikolaev, 1981).

Table 3.5. Livestock number in Turkestan in 1899, thousand head (Anonymous, 1902b)

Regions	Horses	Cattle	Camels	Sheep and goats	Donkeysand mules	Total
Semirechie	637.8	419.8	83.0	4410.0	3.0	5553.6
Samarkand	91.7	183.6	44.6	1069.0	43.0	1431.9
Syrdarya	434.5	516.2	395.0	4754.6	13.1	6113.4
Fergana	38.1	451.2	23.9	1327.7	18.4	1859.3
Transcaspian	123.5	48.0	232.2	3676.7	15.5	4095.9
Total	**1325.6**	**1618.8**	**778.7**	**15238.0**	**93.0**	**19054.1**

ter cold was followed by summer drought, and forage plants had low productivity. Davletov (1977) found, that the following winters in Turkmenia were especially severe: 1899 – 1900, 1903 – 1904 and 1910 – 1911. About 70% of the livestock perished during those years. The people of the Mangyshlak uyezd lost 700 thousand head of livestock in the winter of 1910 – 1911.

Forestry

The Russian administration in Turkestan made the first attempts on nature conservation. The first inventory of forests was conducted and statistical information was collected (Palen, 1910a). General Kaufman called upon the people of Fergana in 1881 for stabilization of moving sands by conservation of vegetation and planting trees. In the Fergana region since 1882 cutting reeds was allowed only by special permission of the administration.

All forests of Turkestan were declared as state property. In 1889 the first forest inventory began, when forest management covered vast areas. The first forest surveyors concluded that forests were destroyed by irregular cuttings, especially tugai. Table 3.6 give data on forests in the beginning of the 20th century .

According to the first classification, forests of Turkestan were divided into four categories: steppe forests, tugai, mountain forests and plantations. So called "steppe forests" were wood land consisting of saxaul and other desert shrubs. Mountain forests usually had low crown density (less than 0.30) and were destroyed by cuttings. Forest plantations occupied a very small area (less than 2,000 ha).

Table 3.6. Forest land in three regions of the Turkestan territory, thousand ha (Palen, 1910a).

Regions	Forest and other wooded land	Forest area	Percentage of forest cover
Syrdarya	51993	14824	28.5
Fergana	13210	10500	7.9
Samarkand	73270	20274	62.6
Total	**128473**	**45598**	**27.6**

The detailed forest survey was conducted mainly in mountain areas where several valuable tree species were common. These data are given in Table 3.7. But the results of this inventory embraced only a part of the forest area.

Burning charcoal was very popular in Turkestan. It could be obtained from 1 cubuc sagene (9.66 m^3) 27 – 30 pood (442 – 449 kg) of charcoal for samovar or 17 – 20 pood (278 – 327 kg) for smithy. Saxaul and archa gave the best charcoal. Making charcoal from growing trees was forbidden by the Russian administration.

The population of Turkestan was allowed to graze animals in all types of forests without any limitation. But according to the opinion of all forest experts, grazing livestock caused great damage to forest stands. Animals grazed the young growth and trampled herbaceous plants. Camels and goats were especially undesirable animals. All attempts of the Russian administration to regulate grazing in forests failed. Russian officials said: "It was impossible either to regulate or to forbid grazing in the forest" (Palen, 1910a).

A new administration started a program of afforestation. In the three regions mentioned above three forestry farms were set up. Their area to-

Table 3.7. Distribution of surveyed forests by dominant tree species, thousand ha (Palen, 1910 a).

Tree species	Region			
	Syrdarya	Fergana	Samarkand	Total
Archa	122.6	412.3	309.0	843.9
Nut-tree	2.4	36.3	-	38.7
Deciduous hrubs	25.4	57.7	2.3	85.4
Mapple, apple tree, hawthorn	20.8	107.4	3.0	131.2
Poplar, birch	5.8	9.4	0.4	15.6
Willow, ashtree, spruce	2.1	14.6	-	16.7
Corc elm	0.1	0.1	-	0.2
Total	**179.2**	**637.8**	**314.7**	**1131.7**

taled 4,422 ha. In the first years of their activity, 274, 000 seedlings of tree species were sold to the local population, 107,278 seedlings were distributed among the local people free. 545 hectares of mountain slopes were covered with forest plantations in the Syrdarya region. Mud flows (sil') were stopped. General Korol'kov ordered establishment of plantations of thuya and beech in the vicinity of Samarkand (Moser, 1885). A part of these plantations has been conserved up to present days.

The Russian administration met a problem of sand drifts menacing the Transcaspian Railway, for example between Merv and Charjui. So, the Russian government established a nursery garden in Farab for growing seedlings of plants for protection of railways against sand drifts. It should be mentioned here that this nursery is still functioning.

Forest surveys were not conducted in the Semirechie region. By rough assessment, the forest area totaled 1,394,000 ha (Palen, 1910a). Mountain forests occupied 208,000 ha; the rest were tugai and steppe forests. The administration of this region introduced in the 90 years of the 19th century several regulations in forestry. Construction of houses and fences from wood were forbidden. Wood could be used only in special cases. Grazing livestock in forests was also forbidden. Cutting wood totaled (m^3):

Years	Growing wood	Dead wood
1905	68,832	188,206
1906	36,486	181,444
1908	51,063	131,177

The detailed forest inventory was not conducted in the Transcaspian region, only small tracts being surveyed. The forest in this region was divided into the following categories: sandy forests, mountain forests, tugai, and protective forests in the Kara-Kala district. Sandy forests occupied the area of 8.74 mln. ha. Especially of great value were mountain forests in which archa, pista chio-tree, maple and other tree species dominated. It was known then, that valuable tree species were common in the Kara-Kala district; for example pear-tree, walnut, plane tree and several medical and food species. The administration of the Transcaspian region adopted several regulations on protection and exploitation of forests but the local people didn't observe them.

The total growing stock of forests in Transcaspian was estimated as 43.47 mln. m^3, proceeding from the assumption that the mean volume of wood totaled 0.5 cubic sagene per dessiatina (4.43 m^3/ha).The age of cutting saxaul was estimated to be 50 years; that for Tamarix species was 5 years. The tariff for wood cutting was 2 roubles per 1 cubic sagene (9.99 m^3). Saxaul was allowed to be cut if the stem diameter near the ground was 2 vershok and more (8.9 cm). Animal grazing was allowed everywhere in the desert. Illegal wood cutting was punished: 1 roubles for one desert acacia, 1 roubles for one pistachio-tree, 10 roubles for one almond-tree, 50 roubles for one walnut or plane tree. Tree cutting was forbidden along the Transcaspian railway in the protected strip, 10 verst in width, 5 verst from each side of the railway. But all measures on conservation of forests and regulation of wood cutting didn't have any considerable effect. Natives cut forests without any permission. Saxaul stands were partially cut along the railway in an strip 40 km in width. In the railway strip between two stations, Repetek and Peski, saxaul was cut on an area of 17,400 ha.

Khiva and Bukhara

We dont have statistical information on the territory of Khiva and Bukhara because these khanates formally didn't belong to Russia. According to the treaty signed in 1873, the khan of Khiva became a vassal of Russia, but Russia didn't interfere in the inner affairs of Khiva. The Khiva khanate occupied the area of 61,734 km², population totaled 700,000 (Anonymous, 1902a). The territory of Khiva consisted of two parts: oasis and desert (sandy, loamy and saline).

Oases were located in the Amudarya delta. Water flowed to the Aral Sea through two main channels: Ul'kun and Yany Su. The river was navigable from Chardzhou to Petroaleksandrovsk. The only steamer named, "Tsaritsa", made the trip in 3 days downstream and 5 days upstream.

Irrigation lands in Khiva totaled about 240,000 ha. The people cultivated wheat, mung bean, sesame, sorgho, lucern, millet, cotton, madder. Potatoes and cucumbers were cultuvated in small quantities. The mean yield of grain crops was 2.25 metric ton/ha, that of sorgho 3.76 metric ton/ha. The oases as a whole produced annually 24,000 metric tons of grain and 6,500 to 9,800 metric tons of cotton. Cericulture was known for a long time, but its productivity was very low because of diseases of caterpillar's bombyx. Livestock numbers totaled (thousand head): horses 100, camels 130, cattle 120, sheep 960, goats 179. The catch of fish totaled 820 metric tons per year. The main food fishes were: *Acipenser schypa, Aspius esosimus*, and *Barbus brachycephalus*.

The Bukhara Khanate was formally independent, but *de facto* it stayed under full control of Russia. The khanate occupied the area of 239, 400 km², its population totaled 2 mln. Roughly 10% of the area was cultivated. Production of cotton totaled 35,000 metric tons, that of raw silk 10,000 metric tons. Production of grain was insufficient, so the Samarkand region supplied Khiva with grain (Anonymous, 1891).

3. 4. Aborigines and new comers in the desert

People of the past

The modern population of Central Asia represents the mixture of two ethnic elements: Turko-Mongolian and Indo-European. During many centuries Turkestan was a scene of bloody fighting between aborigines and new comers. The first reliable information on the population of Central Asia was obtained by a general census in the

Table 3.8. The area and population of Central Asia (Anonymous, 1891, 1902a, 1902b)

Regions	Area, km²	Population	Population density/km²
Russian Turkestan			
Semirechie	394,395	990,107	2.5
Samarkand	68,962	857,848	12.4
Syrdarya	510,292	1,479,548	2.9
Fergana	160,141	1,560,411	9.7
Transcaspian	554,860	372,193	0.7
Total	**1,688,650**	**5,130,106**	
The Khanates			
Bukhara	239,400	2,000,000	8.3
Khiva	61,734	700,000	11.3
Total	**301,134**	**2,700,000**	
Total in Central Asia	**1,989,784**	**7,830,106**	

Russian Empire conducted in 1899 (Table 3.8). The table also gives data on the population of Khiva and Bukhara, where the census was not conducted.

So called "national problem" was settled in different ways before the Bol'shevik revolution in 1917, during the Soviet period and after 1991 when the former Soviet republics proclaimed their independence. The names of nations were also changed. Sometimes statistical information on population was falsified to please political leaders.

Palen (1910b) expressed an official approach on the national problem in Russian Turkestan. Russian scientist Semenov-Tianshansky (1913) supported this point of view. According to official sources the composition of population in Central Asia was as follows.

Turko-Mongolian people represented the main part of population. The Kirghizes (36%) were the most characteristic ethnic group. They were divided into two main groups: steppe Kirghizes (Kazakhs) totaling 29% and mountain Kirghizes (7%). But the name "Kazakh" was not used during the pre-revolutionary time. Nomadic and seminomadic cattle breeding was the main occupation of this group. They lived in yurtas, but a part of them was engaged in cropping at that time.

The Turkmen were also nomads, totaling 5%. They mainly lived in the Transcaspian area, but they didn't have any state system. Some Turkmenian tribes conducted plunder raids on Persian provinces. Elected captains (serdars) were heads of these bands. Turkmenian bands seized prisoners and sold them on slave markets. Slave trade was forbidden by the Russian administration.

An attempt was made to revive the slave trade in Tajikistan in 1992. Mr. E. Rokhmonov, the head of the State told about capture of women and children. Local guerrilla bands conducting a civil war led them to Afghanistan and exchanged captives for weapons (Babakhanov, 1992).

The Uzbeks and the Kipchaks (19%) also belonged to the Turko-Mongolian group. They mainly populated Samarkand, but also lived in Fergana and in Syrdarya. The name "Uzbek" appeared in the 14th century rather as a political but not an ethnic symbol during the rule of an Uzbek khan. As known, he was a zealous Moslem and had done much for dissemination of Islam.

Bartol'd (1914) explained the invasion by Uzbeks of Central Asia in the following way. Barbarian tribes conquered Turkestan in the 14th century, they came from the territory of the Golden Horde. These tribes were called Kirghizes in the north and Uzbeks in the south.The Uzbeks formed two khanates – Bukhara and Khiva. The Kokand khanate was formed later. The Uzbeks settled, step by step. In the mind of subject tribes, the Uzbeks were a superior race, the master and the boss!..

A considerable part of the population (24%) belonged to Sarts, a people of unknown origin. Palen (1911) issued the following idea about their origin. According to the notes of the Chinese traveler Jen Kyan (126 B.C.), the steppe part of Turkestan was populated by nomadic Turk tribes. Settled Aryan tribes lived in valleys of the Zeravshan, the Chirchik and the Angren. Irrigation agriculture was developed there in that time. Turk nomads under the thrusts of Mongolian hords began to press the Aryans, then Turks mixed with them. That resulted in the appearance of a new population in Turkestan – the Sarts.

Semenov-Tianshansky (1913) said about the Indian origin of the word "Sart", that meant "merchant". The Sarts used their own language (Sart-Tili), different from Uzbek. Rickmers (1913) said that Sarts were town dwellers of Bukhara, Samarkand and Khiva, they were not nomads, not Jews, not dwellers of mountains. Sometime the Russians used the name "Sart" with scorn, like Americans said "Nigger".

Philoprogenitiveness was a characteristic feature of Sarts. They said: *"Many children – bazar, no children – mazar"*. By the way, numerous families were common to other Moslem groups. This tradition has been conserved up to present days in Central Asia.

From Indo-European groups, the Tajiks were the most numerous (9.5%). They lived in the mountains of Samarkand and Fergana regions. The Uzbek conquerors forced them out of their homes into mountain areas.

Many other small national groups were also registered in Central Asia during the pre-revolutionary period. To them belonged Bukhara Jews,

Taranchis, Dungane, Gipsies, etc. Bukhara Jews were an oppressed national minority (Semenov-Tianshansky, 1913). Bukhara was their religious and spiritual center. Very likely they were offspring of those Jews who had been taken prisoners by Iranians and Assyrians during the war with Israeli and Judaic kingdoms.

Bukhara Jews spoke a dialect of Persian. They resided in special blocks. They were allowed to wear only gray khalats girdled by ropes, they couldn't ride horses (only donkeys !), it was forbidden for them to leave their houses after sunset etc.Only after establishment of the Ak Padishakh's power did the Jews become equal in their rights with other groups.

Under the name Taranchis were known the so called "Alian people", who came from China. They resided in the Semirechie region and were good farmers. Some of them were Turks; the southern group was close to Sarts.

Dungane were Chinese people, who adopted Islam. In China they had the name of "junior people" (syao jao). The Chinese called themselves the "senior population" (da jao).

The Russian population in Turkestan was not numerous, totaling 4.5%. These included military, administrative personal and merchants. Sixty Russian villages, including 29 stanitsa of Kozaks, were set up in the vicinity of Vernyi. Table 3.9 gives information on occupation of the people in Central Asia.

The growth of population in Central Asia (% per year) is characterized by the following figures: Semirechie – 0.7, Syrdarya – 1.4, Fergana – 2.2,

Table 3.9. Occupation of the people in Central Asia, 1911, in percentage (Palen, 1911).

Nations	Semirechie	Samarkand	Syrdarya	Fergana	Transcaspian
Russians A	44.26	5.56	18.54	1.35	.30
L	0.40	0.03	0.12	1.06	0.13
O	55.34	94.41	81.34	98.31	98.57
Kirghizes A	41.14	19.15	45.48	75.00	0.10
L	54.23	78.40	46.97	17.96	96.46
O	4.63	2.45	7.55	7.04	3.44
Sarts A	19.33	44.54	57.55	60.00	-
L	1.20	0.78	1.62	0.53	-
O	79.47	54.68	40.83	39.47	-
Uzbeks A	-	82.07	69.04	44.72	-
L	-	2.16	1.72	0.48	-
O	-	15.77	29.24	54.80	-
Tajiks A	-	37.81	46.42	44.72	-
L	-	1.22	1.22	0.48	-
O	-	60.97	60.97	54.80	-
Turkmen A	-	-	-	-	69.26
L	-	-	-	-	16.67
O	-	-	-	-	14.07
Kara- Kalpaks A	–	-	85.66	89.60	-
L	-	-	9.11	0.03	-
O	-	-	5.23	10.37	-
Taranchis A	68.53	-	-	-	-
L	0.81	-	-	-	-
O	30.66	-	-	-	-
Dungane A	46.90	-	-	-	-
L	0.58	-	-	-	-
O	52.52	-	-	-	-

Note : A – agriculture, L – livestock raising, O – others.

Transcaspian – 1.4. The biggest towns of the regions were: Tashkent (189.7 thousand people), Kokand (11.8 thousand), Namangan (70.3 thousand), Samarkand (69.3 thousand), Andizhan (72.9 thousand).

Bukhara had a population of about 60 thousand. Travelers of the past usually overerestimated the number of the people living in this town. Bukhara had 50 markets (bazar) always crouded by people (only by males). Foreigners were sure that only half of the population was in the bazars, but really near 100% of males "pushed one other in markets!".. As known, the bazar, for oriental people, always was a kind of club where people exchanged news and obtained various kind of information. Bazars in Central Asia never lost this function.

In the Bukhara Khanate, Uzbeks were the dominant nation. The other national groups included: Kirghizes, Sarts, Tajiks, Jews, Afghans, Armenians, Gypsies, Hindus etc.

According to Moser (1885), in the Khiva Khanate the national composition was as follows: Sarts, Uzbeks and other settled groups numbered – 43%, Kirghizes –13%, Kara-Kalpaks – 15%, Turkmen – 29%.

The socio-ecological condition of the people in Central Asia during the pre-revolutionary period could be described as follows. The local people lived in reed huts coated with clay. Prosperous people lived in houses made of saman (Lessar, 1884). Foreign travelers called these houses "loess architecture".

Houses of Sarts in Bukhara had, as a rule, a small pond in the yard (khaos). Water for cooking and drinking was taken from this pond. Ritual ablutions were also performed in the same pond. Rickmers (1913) said that the size of Sart houses was about 30 m^2. Such living space was very limited if one takes into consideration the large families of natives.

Grulev (1901) gave a description of a house of a prosperous Turkmen. The house was constructed on the principle "my house is my castle". The yard was surrounded by saman wall, there was only one entrance. Female apartments were located separately, according to Moslem traditions. The houses of Sarts and Uzbeks had the same design.

The terrible unsanitary conditions of the people should be mentioned here. People never washed, except the ritual ablutions prescribed by Islam. Women washed their hair with sour milk. Cloths were never laundered either. They were worn out and then thrown away.

In towns of Central As is populated by Sarts and Uzbeks, public baths were known from ancient times. In Bukhara alone there were 16 baths (Anonymous, 1891). But these baths were mainly used by rich people, the poor didn't visit baths.

Unsanitary conditions and poverty stimulated the distribution of many diseases. Epidemics of small pox, cholera, malaria and spotted fever were common in the region... Infant mortality was high, nearly 50% of infants died before reaching the age of 1 year. Twenty thousand people died from malaria in 1895 – 1896 in Merv uyezd of Trans-caspia, that made 18% of the population. Cholera was registered in Ashkhabad in 1892 – 1893 (Kiyatkhanov & Tesler, 1984). Trachoma was a real scourge of nomads, especially Turkmen. Tuberculosis was also a very common disease among people in the desert.

Rishta was also very common disease; it was also known as "sart diseas" or "cotton thread". Rishta was caused by a tropical worm which got under the human skin. The worm could reach two or three feet long. Even the emir of Bukhara was infected by rishta. Local quacks and barbers could extract the worm by cutting the human skin and winding the worm on a match.

Rickmers (1913) found many people infected by leprosy. Leprous with their families lived in special blocks. Local people avoided contact them. Lepers often made raids to fields and gardens where they took vegetables and fruits. After a leper touched something, this object was considered unclean. Lepers were allowed to be beaten by stones from a distance.

The status of woman in a feudal society was very low. Uzber and Tajik women covered their faces with a special veil (paradzha). Kirgiz, Kazakh and Turkmen women didn't cover their faces and their role in family was a little better. Bat "the sterner sex" forced them to observe Shariat and young girls especially were suffered. They could be sold to husbands, 12–13 years old. According to a Turkmen saying, a young girl is mature for mar-

Weeps of Moslem wife

Less than the dust beneath the chariot wheel
Less than the rust never stained the sword
Less than the trust thou hast in me, my lord
Even less than these!
Less than the weed that grows beside the door,
Less than the speed of hours spent far from thee,
Less than the need thou hast in life of me
Even less am I!
Since I am my lord, am nothing unto thee,
See here the sword, I make it keen and bright
Love's last reward – death come to me to-night.
Fare well, Zahirudin !

riage if she couldn't be knocked down by papakha. Zwemer & Zwemer (1926) described the complaints of Moslem wife. These lines are worth to cite them below:

Rickmers (1913) described an accident which characterized the morals of Moslems in Russian Turkestan. Once two Uzbek women (certainly accompanied by men!) were driving in an arba hauled by a donkey. Their faces were covered by parandzha. While crossing a big canal, the arba turned over because the wheel fell into a pit. The canal was not so deep and the driver saved themselves without great difficulty. But the unfortunate women drowned!.. The women could also escape, but they were afraid to bare their faces while floundering in the water. They prefer to die before violating Adat.

Rickmers (1913) made a very interesting observation on the diet of the people in Russian Turkestan. Neither Sarts nor Uzbeks had fixed hours for meals. All food cooked was eaten at once without stopping. Some days, for example during festivals, the food could be plentiful, during other days the people satisfied themselves with tea, lepeshka and dried fruits. It was a common food of Sarts. Meat was seldom consumed.

Yuferev (1925) studied the diet of cotton growers in the Fergana region. His study concerned the life of the rural people before the First World War. The food of Sarts in this region was poor in protein and fat, but they consumed much carbohydrates. That made the food coarse and bulky. Cotton growers who possessed small land holdings couldn't make both ends meet. They didn't accumulate any capital and often ruined themselves.

But we could see one positive ecological feature of agriculture in Russian Turkestan. At that time agriculture was ecologically pure branch of the national economy. Mineral fertilizers, herbicides, pesticides and other chemicals were not used in agriculture.

Alcoholism, a social scourge in the modern society of Central Asia, was not common in those days. But this "tabu" was broken from time to time. Grulev (1901) found that Turkmen drank vodka secreatly. Jews or Armenians were suppliers of alcohol. According to observations of Donavan (1883), the Turkmen in Merv also consumed vodka, bur that mainly concerned serdars and participants of alaman. But all these cases were rather exceptions to the rule.

Teriak and has hish were used by natives. These drugs were delivered from Iran and Afghanistan, despite the tsarist administration having forbidden the trade of drugs.

Living conditions of nomads in the desert were not so pleasant. Nomads lived in yurtas. In the center of the yurta the fire made of saxaul was a source of heating, food was also cooked on it. Often two families lived in one yurta – a family of elders and a family of a married son who couldn't create their own hearth.

A new commonwealth of "the Socialist nations"

A general demarcation of Central Asian republics was conducted in 1924 – 1925 on the direct instructions of J. Stalin. The future dictator was afraid of the potential unity of large Turk nation. It would be more suitable to divide this nation to separate tribes and to label them "nations".

Other groups who contradicted this concept disappeared, not *de facto* but *de jure*. The Sarts, a group numbering about l million before the Bol'shevik revolution, disappeared. In the census the most of the Sarts were registered as Uzbeks because Uzbeks were always, for them, "the higher race". And the temptation was great to improve their rating! You see, the people of those days didn't have any papers.

Reklyu (1892) noted that situation. She indicated, that the number of Uzbeks was increasing on account of other national groups joining them. Taranchis and Dungane also disappeared. Many Jews adopted Islam and were registered as local groups.

After the general demarcation, the administrative boundaries of Turkestan were drastically changed. Instead of the former limits, which often coincided with physical features, new, more winding lines cut a political map of Central Asia. Demarcation was conducted on a national principle – each ethnic group received its own territory. So, that time a basis was created for future national conflicts. The territory of Russian Turkestan was divided before on two other principles: dynastic (for example, the State of the Timurids) or domain (for example, the Khanate of Bukhara). Ak Padishakh didn't divide the territory on an ethnic groups.

The new demarcation immediately caused several national conflicts because many ethnic groups were mixed, not separated. Table 3.10 gives the distribution of the population in Central Asia after the general demarcation.

After the general demarcation many people changed their residence. For example, many Uzbeks from Southern Kazakhstan and Kirgiziya came to Uzbekistan, the Mangyshlak Turkmen migrated to Turkmeniya, the Kazakhs from Uzbekistan migrated to Kazakhstan, etc. (Central Asia, 1968).

The change of the population in the later period of time is characterized Table 3.11.

Pripiska was made in the population census. Germans were not registered in Kazakhstan in 1970! 1,398 thousand people were included into the category "others", that made 10.2% of the population in the republic. These figures included mainly Germans who were deported to Kazakhstan in 1941 from the Autonomous Republic of the Volga Germans. They demanded in 1990 – 1992 the right to re-establish their autonomy but in vain!

In 1991 the last statistical information on the USSR economy was published. The number of population in Central Asia is given below (Table 3.12).

In this table the national composition of the population is not indicated. It was the last pripiska in population census made in the Soviet Union. But that couldn't prevent the national conflicts in the former Soviet Republics, especially the civil war in Tajikistan.

The growth of the population in Central Asia was high, especially if it is compared with Russia (Table 3.13). The growth of population in this region was 10 times heigher than in Russia and 20 times heigher than in Byelorussia. That fact testifies to the policy of genocide to Slav nations conducted by the central government of the USSR. Table 3.14 gives information the level of education in Central Asia and Russia.

But the table gives formal criteria. One should take into consideration that the level of training cadre in Central Asia was very low. Many people were trained in local languages, students paid bribes, and practically 100% of young people joining universities got diploma. Quantitative criteria in education played the most important role in the national policy of Central Asian Republics. During the Soviet period the natives mastered the Russian language. The percentage of the people speaking Russian fluently is given below: Uzbeks – 49.3, Kazakhs – 52.2, Tajiks – 29.6, Turkmen – 25.4, Kirghizes – 29.4.

The mean size of the family in 1989 was: Russia – 3.5 people, Uzbekistan – 5.5, Kirghizstan – 4.7, Turkmenistan – 5.6, Tajikistan – 6.6. There were two main reasons for the population growth

Table 3.10. Number and the national composition of population of Central Asia in 1926 (Central Asia in figures, 1931).

Nations	Uzbek SSR	Tajik SSR	Turkmen SSR	Kirghiz SSR	Kara-Kalpak autonomic region	Total
Uzbeks N,	32,251,212	255,567	109,971	109,776	84,099	3,779,625
%	76.2	24.7	10.1	11.1	27.6	49.6
Turkmen N,	21,797	4,157	719,722	3.0	9,686	755,435
%	0.5	0.4	60.1	-	3.0	9.9
Kirghizes N	72,185	13,356	-	661,171	277	746,989
%	1.7	1.3	-	66.4	0.9	9.8
Tajiks N	231,332	736,338	566	2,667	67	970,970
%	5.4	71.2	-	0.2	-	12.7
Kara-Kalpaks N	26,653	-	1,537	19	116,125	144,244
%	0.6	-	0.1	-	*38.2*	1.9
Others N	195,047	8,212	14,589	13,134	86,790	317,767
Total %	4.6	0.8	1.3	1.3	28.3	4.2
Russians N	237,708	8,820	75,357	116,436	4,924	443,245
%	5.6	0.9	7.2	11.6	1.3	5.8
Other new comers, N	229,623	7,673	128,626	95,403	2,579	463,904
%	5.4	0.7	12.2	9.5	0.7	6.1
Total N	**4,239,467**	**1,034,123**	**1,045,433**	**998,607**	**304,547**	**7,622,179**
%	**100**	**100**	**100**	**100**	**100**	**100**

Table 3.11. Number and the composition of the population in Central Asian Republics in 1970, in thousand/ % (National Economy of the USSR, 1972).

Nations	Kazakh SSR	Kirghiz SSR	Uzbek SSR	Turkmen SSR	Tajik SSR	Total
Kazakhs		22/0.8	476/4.0	69/3.2	8/0.3	575/3.0
Kirghizes		1285/43.8	111/0.9	-/-	35/1.2	1431/7.5
Uzbeks		333/11.3	725/65.5	179/8.3	666/23.0	8903/46.4
Turkmen		-/-	71/0.6	1417/65.6	11/0.4	1499/7.8
Tajiks		22/0.7	449/3.8	-/-	1630/56.2	2101/10.9
Russians		856/29.2	1475/12.5	313/14.5	344/11.9	2986/15.6
Ukrainians		120/4.1	112/0.9	35/1.6	32/1.1	229/1.6
Karakalpaks		-/-	230/2.0	-/-	-/-	230/1.2
Tatars		69/2.4	-/-	36/1.7	71/2.4	176/0.9
Koreans		-/-	148/1.3	-/-	-/-	148/0.8
Jews		-/-	103/0.9	-/-	-/-	103/0.5
Armenians		-/-	-/-	23/1.1	-/-	23/0.1
Others		201/6.9	328/2.7	87/4.0	103/3.5	719/3.7
Total		**2908/100**	**11226/100**	**2159/100**	**2900/100**	**19193/100**

Table 3.12. Population of Central Asia in 1991, in thousands (National Economy of the USSR, 1991).

Republic	1979	1989	1991
Kazakhstan	16,484	16,536	169,674
Kirghizstan	3,529	4,290	4,484
Uzbekistan	15,391	19,905	21,207
Turkmenistan	2,759	3,534	3,809
Tajikistan	3,801	5,109	5,370
Total	**40,160**	**49,374**	**51,834**

in Central Asia: Moslem traditions and better medical service compared with the pre-revolutionary period.

The expected length of life is given below (Table 3.15) The expected length of life of women is lower in Central Asia than in Russia. That can be explained by the lower status of women in Moslem society. Oases in the Fergana valley had the most dense population, numbering 500 persons per km^2 (Central Asia, 1968). In desert areas, population centers were very rare – 3 – 10 per 1,000 km^2. Their size was very small, numbering several dozens persons. Exploitation of mineral resources, especially oil and gas was the reason for setting up new settlements and towns, like Shatlyk, Gas-Akcha, Neftezavodsk, etc. The distribution of population in Central Asia was always irregular. The main part of the population was concentrated in oases, along the railway, in river valleys and around water sources.

3.5. We shall even control the drought !

"We shall even control the drought"– that was a slogan advanced by J. Stalin during the last years of his life. Stalin, "the great leader and teacher of the Soviet people" tried to transform not only human society but also to remake nature. According to his ideological concept, nature was a passive physical body which should be transformed for the benefit of the Soviet people. The concept of remaking nature became a paradigm of Soviet leaders who wanted to realize this idea first of all in agriculture. This paradigm was supported by apologists of a new ideology in scientific circles, for example by T. Lysenko.

So an ideological approach was the first reason why new canals should be constructed and protective forest belts should be established. The second reason was political. Soviet people needed slogans for reconstruction of the national economy destroyed during the Second World War. Economic considerations were put in third place!

Stalin died in 1953. His followers continued the policy of nature remaking. N. Khrushchev ordered the ploughing of vast areas of virgin land in Kazakhstan. That resulted in severe wind erosion, especially in dry years. The theory of V. Vernadsky about noosphere was used for speculations. As known, noosphere was a kind of an ideal society in which people would live in full harmony with nature...

The project of the Main Turkmen Canal

The newspaper "Pravda" published on 12 September 1950 the decree of the Soviet government about construction of the Main Turkmen Canal (MTC). The decree said: "MTC, 100 km in

Table 3.13. Birth rate, death rate and population growth, per 1000 people (National Economy of the USSR, 1991).

Republics/ Years	Born		Died		Growth	
	1980	1990	1980	1990	1980	1990
Russia	15.9	13.4	11.0	11.2	4.9	2.2
Kazakhstan	25.1	21.7	8.0	7.7	17.1	14.0
Uzbekistan	33.9	33.7	7.5	6.1	26.4	27.6
Kirghizstan	29.6	29.3	8.4	7.0	21.2	22.3
Turkmenistan	33.4	34.2	8.3	7.0	26.0	27.2
Tajikistan	37.0	38.8	8.6	6.2	28.9	32.6

Table 3.14. Number of people with university education and incomplete higher education, per 1,000 people (National Economy of the USSR, 1991).

Republics/ Years	University education		Incomplete heigher education	
	197 9	1989	1979	1989
Russia	77	113	348	483
Kazkhstan	73	98	440	511
Uzbekistan	70	92	443	577
Kirghizstan	68	94	397	564
Turkmenistan	64	83	392	568
Tajikistan	60	75	365	551

length will be constructed from Takhia Tash on the Amudarya to waterless areas of the Caspian lowland. MTC will run round the Sarykamysh depression, then it will cross the Karakum desert and will follow the ancient channel of the Uzboi". It was planned to irrigate 1.3 mln. ha of virgin lands. MTC should be constructed during 1951 – 1957. The decree gave impetus to an hysterical campaign on nature remaking in the USSR. Party leaders and scientists published many papers and the mass media provided information on "the great project of communism".

Mr. Batyrov (1950), the former First Secretary of the Communist Party of Turkmenistan, said: "During 1934 – 1939 the number of livestock in the Republic increased 3 times because of development of new tracts of rangelands. During the last decade the livestock again increased 3 times". In his opinion, the construction of MTC would improve these figures. But his follower, Mr. Ovezov (1967), indicated many bottlenecks in development of livestock raising. In his opinion, the low level of range management was the main reason for reduction of the number of livestock during the same period...

Mr. Yusupov (1950), the USSR Minister of Cotton Industry, said that MTC would stabilize the flow of the Amydarya, especially in Kara-Kalpakia and Khorezm region where the length of earth dams totaled 880 km. Kovda (1950a) issued the opinion that cotton production in the zone of MTC would total 4.0 metric tons per hectare. Mr. Bovin (1950), the former Forest Minister of the USSR, promised to create "the Green Belt of Karakum".

An idea about the construction of a canal from the Amydarya to the Caspian Sea was issued a long before. Turkmen Khodzhanepes visited Saint Petersburg and had an audience by Peter the Great. He assured the tsar that it was possible to turn the Amydarya flow to the Caspian Sea. That could open a new commercial way to India (Ivanin, 1873). The tsar ordered Prince Bekovich-Cherkassky to collect information on this subject during his expedition to Khiva. Bekovich-Cherkassky was bestially murdered and the poblem was not solved (Moser, 1885).

During the Soviet period several versions of the Karakum Canal were proposed but none of them was realized (Figure 3.8).

The project of MTC was abandoned after Stalin's death. We don't know who was the initiator of this decision because all was done behind closed doors. A report on the role of L. Beriya in solving this problem was published later (Leskov, 1993). The author of this publication interviewed B. Vainstein, retired colonel of the KGB. The colonel highly esteemed the role of Beriya in the Soviet economy. He said, that

Table 3.15. The expected length of life, per 1,000 people (National Economy of the USSR, 1991)

Republics	1979 - 1980			1990		
	Total	Men	Women	Total	Men	Women
Russia	67.5	61.5	73.0	69.3	63.9	74.4
Kazkhstan	67.2	62.5	70.5	69.2	64.3	73.3
Uzbekistan	67.6	64.0	70.7	69.5	66.2	72.6
Kirghizstan	66.0	61.6	70.0	68.8	64.5	72.5
Turkmenistan	64.6	61.6	67.8	66.4	62.9	69.7
Tajikistan	66.3	63.7	68.6	69.6	67.0	72.1

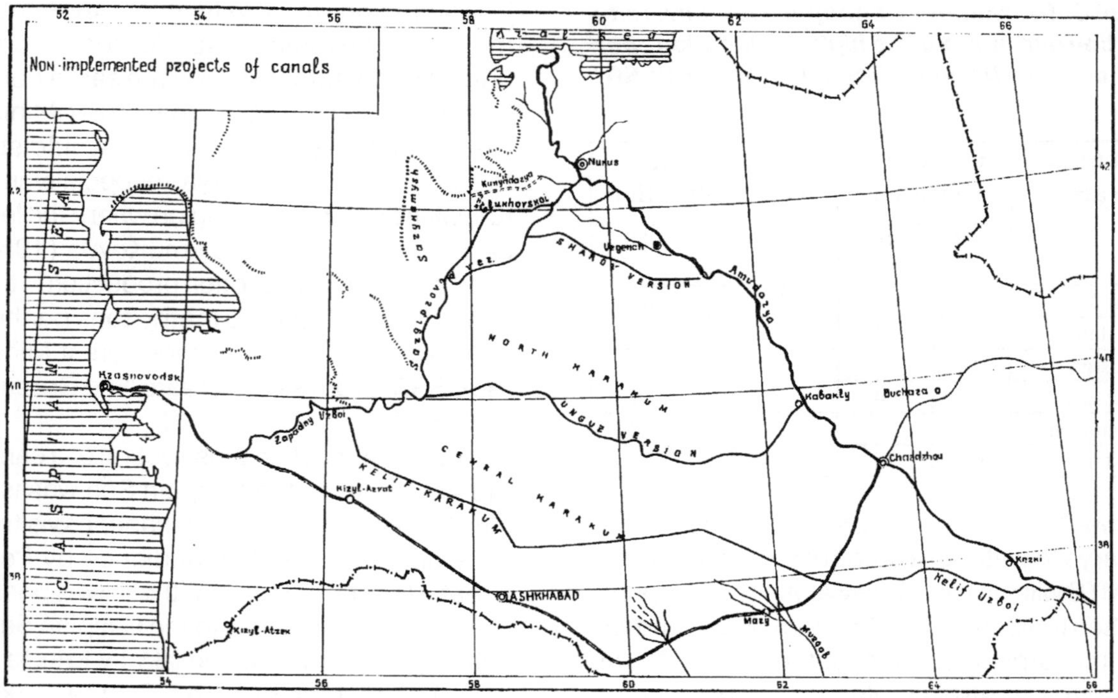

Figure 3.8. Non implemened projects of the canal in Central Asia.

Beriya was sure that such projects as MTC were beyond the State's strength. Beriya argued against many such projects including MTC and the Volga – Don canal.

Transfer of river flow from Siberia

The idea of watering Central Asia by Siberian rivers came to people's mind long before the epoch of L.Brezhnev. In 1869 Russian agronomist Ya. Demchenko presented a proposal to the Russian Imperial Geographical Society about the transfer of Siberian rivers to Central Asia.

During the Soviet period the ideas of remaking nature were issued top-down and were supported by "the broad mass of working people". For example, Fedorovich (1950) said: *"It is becoming more and more clear that in the near future we shall start a grandiose project on transfering to Kazakhstan and Central Asia those waters from Siberia which flow useless to the basin of the Arctic Sea".*

New political leaders, like Brezhnev needed new confirmations of the correctness of *"the Party's line" during the epoch of the "developed socialism". The 25th Congress of the KPSS made the "epoch making decision":* "To conduct the research works and to develop a project on partial transfer of the Northern and Siberian river flow to Central Asia, Kazakhstan and to the basin of the Volga" (Materials ...,1976). A new campaign in the mass media began. Conciliators from science called a special meeting in Tashkent in 1978. Presidents of the Academies of Sciences of Central Asian republics and many scientists and specialists attended the meeting (Ziyatdulaev, 1979).

Schemers and scientific time-servers developed a scientific justification of the project. It was as follows. Central Asia produced in 1976 nearly 100% of the cotton in the USSR, 100% of kenaf, 78% of raw silk, 60% of karakul, 16% of wool, etc. Irrigated farm lands occupied in Central Asia 7.2 mln. ha. During 1970 – 1976 about 1 mln. ha of new land was developed. The transfer of water would provide irrigation for 15 mln. hectares, including 1.6 mln. hectares on the first stage of the project. During 1974 – 1975 some 200,000 ha of land were not irrigated because of a scarcity of water. The route of the canal is given in Figure 3.9.

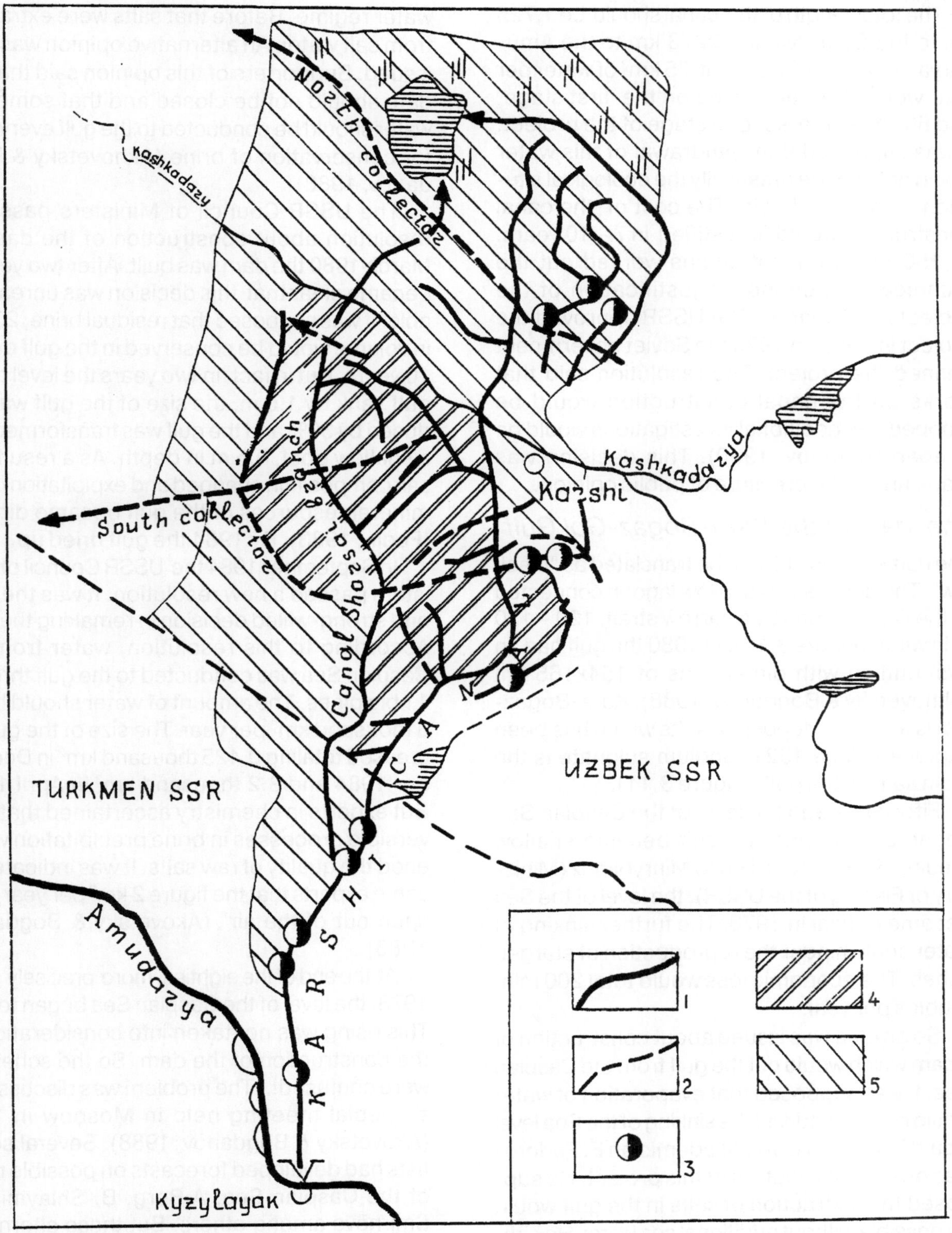

Figure 3.9. The areas of possible irrigation on the account of the Siberian river's water: 1 – dams under projects, 2 – pumping stations under projects, 3 – the main canals, 4 – the branch canals, 5 – areas for irrigation and for rangeland watering (Ziatdulaev, 1979).

The total length of the canal should be 1,725 km to the Syrdarya and 2273 km to the Amudarya. It was supposed that 25 km^3 of water per year would be transferred on the first stage, and 60 km^3 in the second stage of the project. It was supposed that withdrawal of this water would not change drastically the ecological situation in the Ob' basin. The cost of the canal construction would be justified in 7 – 10 years.

150 research institutions worked out the technical and economic justification of the project in 13 years!.. The USSR approved the project in 1983. In 1986 the Soviet government banned the project. The resolution said that works on the canal construction would be stopped and additional investigations would be needed (Antonov, 1990). This decision was made under the pressure of public opinion.

The dam of the Kara-Bogaz-Gol Gulf

The name of this gulf can be translated as "black jaw". The gulf is a vast shallow lagoon connected to the Caspian Sea by a narrow strait, 120 – 130 m in width (Figure 3.10). In 1980 the gulf had an oval outline with dimentions of 154x165 km (Akhovetsky & Bogdanov, 1988). Kara-Bogaz-Gol is a unique deposit of salts which has been exploited since 1921. Sodium sulphate is the main salt of the gulf (Figure 3.11).

Fifty years ago the level of the Caspian Sea began sinking and the gulf became shallow (Fedin, 1983). According to Minrybkhoz (Ministery of Fishing of the USSR), the level of the Sea became critical in 1978. The further sinking of water could hinder the reproduction of sturgen fishes. The economic loss would total 200 mln. roubles per year.

So an idea was issued about construction of a dam which would cut the gulf from the Caspian Sea. It was supposed that evaporation of water would be reduced and the sinking of the Sea level would become slower. Academician E. Fedorov was one of the initiators of this project. He supposed that extraction of salts in the gulf would be possible without additional sea water. He himself and a special commission headed by him came to the conclusion that ancient deposits of dry salts could be exploited by the change of the water regime. Before that salts were extracted from salt water. An alternative opinion was also issued. Supporters of this opinion said that the dam should not be closed and that some sea water should be conducted to the gulf every year for conservation of brine (Akhovetsky & Bogdanov, 1988).

The USSR Council of Ministers passed a resolution about construction of the dam. In March 1980 the dam was built. After two years it became clear that this decision was unreasonable. It was supposed that residual brine, 20 km^3 in volume would be conserved in the gulf during 10 years. But in fact, in two years the level of the gulf sank by 1.5 m, the size of the gulf was reduced by 30% and the gulf was transformed into a shallow pool, 0.5 m in depth. As a result, the salt composition changed and exploitation of the mineral resources of the gulf became difficult (Fedin, 1983). By 1984 the gulf dried up.

In September 1984 the USSR Council of Ministers passed a new resolution. It was the second strong-willed decision in remaking the gulf! According to this resolution, water from the Caspian Sea was conducted to the gulf through 11 big pipes. The amount of water should be 2–3 thousand km^3 per year. The size of the gulf increased totaling 0.425 thousand km^2 in December 1984 and 3.2 thousand km^3 in April 1985. But experts in chemistry ascertained that irreversible processes in brine precipitation worsened the quality of raw salts. It was indicated by some experts that the figure 2 km^3 per year "was spun out of the air" (Akovetsky & Bogdanov, 1988).

At the end of the eighties, more precisely since 1978, the level of the Caspian Sea began to rise. This rising was not taken into consideration by the construction of the dam. So the schemers were confused!.. The problem was discussed in a special meeting held in Moscow in 1988 (Akovetsky & Bogdanov, 1988). Several scientists had developed forecasts on possible rising of the Caspian Sea, A.Berg, B. Shlaymin, K. Smirnova among others. But these alternative hypothesis were not taken into consideration by state schemers.

The same state schemers also did not assess the ecological sequences of the problem. After

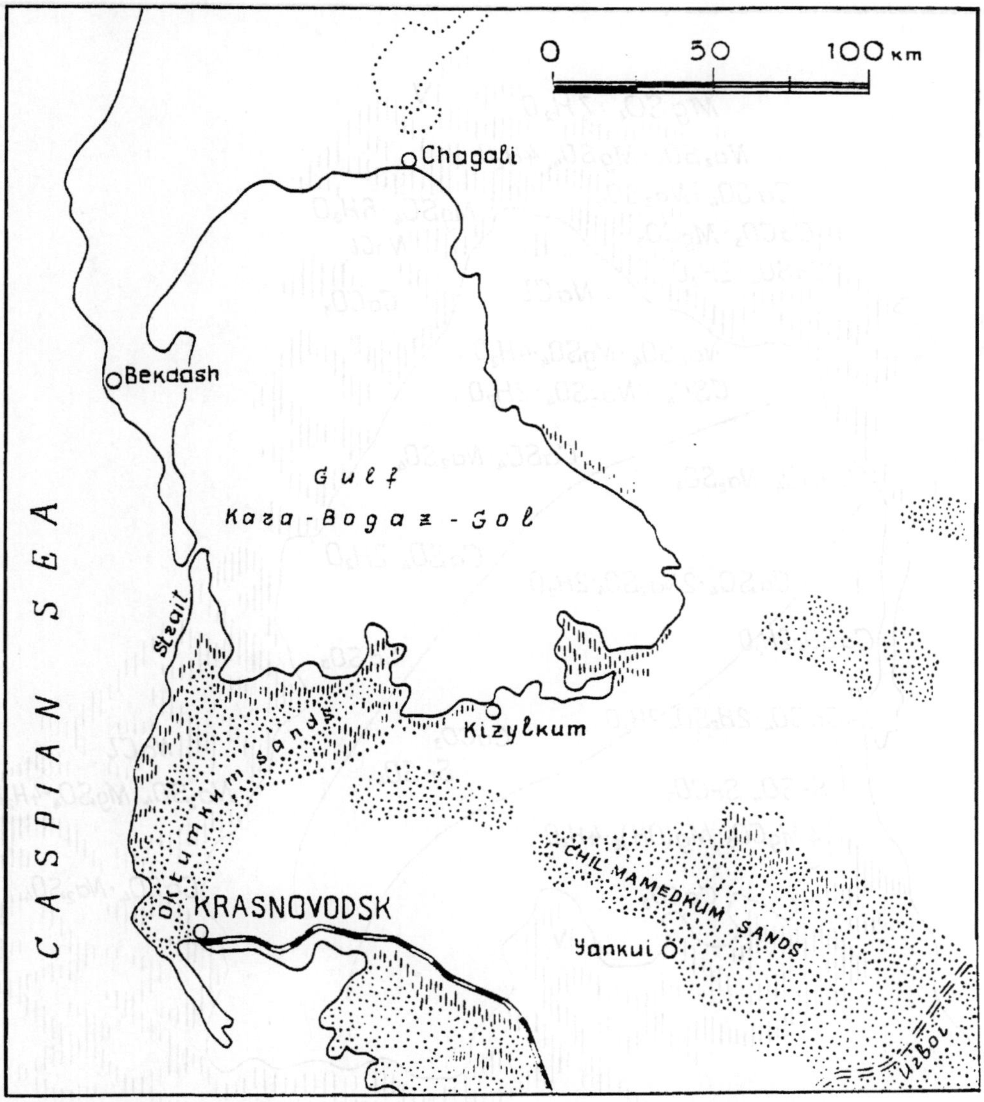

Figure 3.10. The map of the Kara-Bogaz-Gol (Akhovetsky & Bog-danov, 1988)

the daming, this process became intensive. Sulphate storms and whirlwinds transported great amount of salts for hundreds of kilometers. That caused desertification of the bordering areas in Turkmenistan.

Some scientists understood the real situation. But for them it was not safe to continue discussion after the approval of the project by Brezhnev himself! Discussion on the problem was continued during the perestroika period, proclaimed by M. Gorbachev.

But political events which led to disintegration of the USSR determined the fortune of the Kara–Bogaz–Gol. While political storms raged, the life of the Caspian ecosystem continued and the water level was rising. Investigations of Babaev & Laptukhov (1992) couldn't explain why the Caspian Sea was rising. In March 1992 the dam was destroyed by the instruction of S. Niyazov, the president of Turkmenistan (Bushev, 1992a). It was the third strong-willed decision in remaking the gulf!

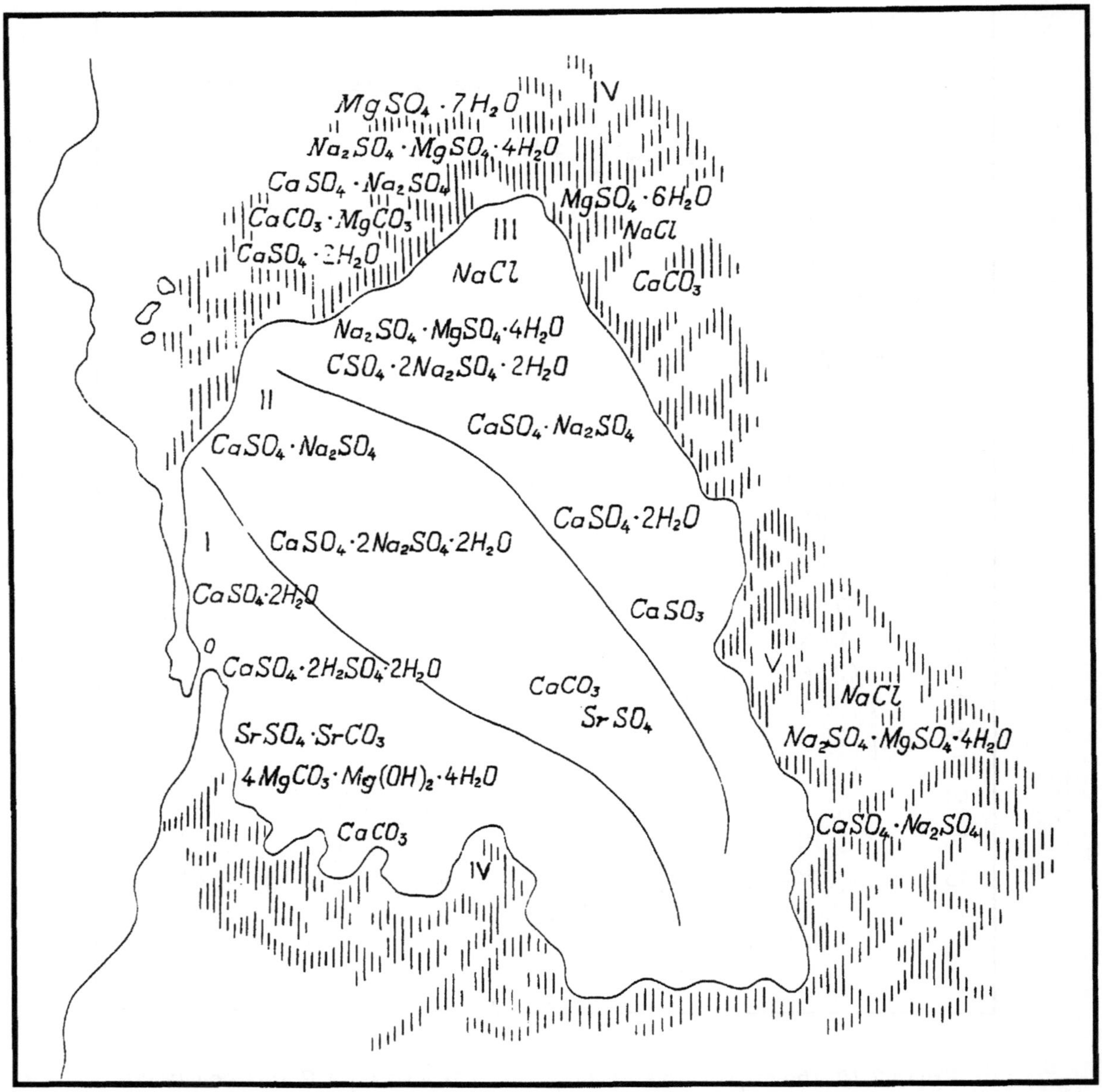

Figure 3.11. Salt depositis in the Kara-Bogaz-Gol (Akhovetsky & Bogdanov, 1988).

The rise of the Caspian Sea contnued untill 1997. The economic loss caused by flooding totaled in 1991 400 mln. roubles (Seiitkurbanov, 1991). Above that, the process of contamination of the Caspian Sea began. Chemical plants of Cheleken and Nebit Dag (Turkmenistan) conducted to the Caspian Sea about 200 mln. m^3 of toxic drainage waters. They contained more than 200 g/l of salts. The content of strontium totaled 300 – 500 mg/l. The water of the Krasnovodsk Bay was contaminated by phenol and other chemicals.

In 1992 an international organization on the problem of the Caspian Sea was established. The members were: Russia, Azerbaijan, Turkmenistan, Kazakhstan and Iran. This organization will study the problem of the Caspian Sea, including fisheries.

3.6. Conclusion

In more than 2,000 years the natural ecosystems of Central Asia have experienced the influence of man (Figure 3.12). During the pre-Mongol invasion period, this influence was comparatively low because of the low level of technology and sparse population. Wood cutting and exploitation of grazing lands were irregular, and some times they were based on tribal traditions.

The invasion of Mongol hordes changed this situation. The Mongols displayed some of the less desirable features of human character. War desertification destroyed flourishing oases of Central Asia. That was not a collateral effect of land reclamation but an intentional destruction of dams, cities and cultural objects. Some oases destroyed by Mongol conquerors have not yet been restored.

During the latest epochs, many rulers tried to glorify their names by construction or restoration of irrigation systems. Surely this activity was not so grandiose as in present days. Surely, in those remote times the scales of nature remaking couldn't be so grandiose as in our days.

The Karakum Canal was a good example of Stalin's ideas about remaking nature. The project of water transfer from Siberia to Central Asia demonstrated the concept of "developed soci-alism" proclaimed by Brezhnev. Gorbachev postponed the project during the perestroika period because of a new ideological concept.

Construction of the Karakum Canal was proclaimed by leaders of the USSR as a great historical event in the life of people of Central Asia. But the grandeur of this project should be considered rather in its negative consequencies – desertification. The people realized that too late!

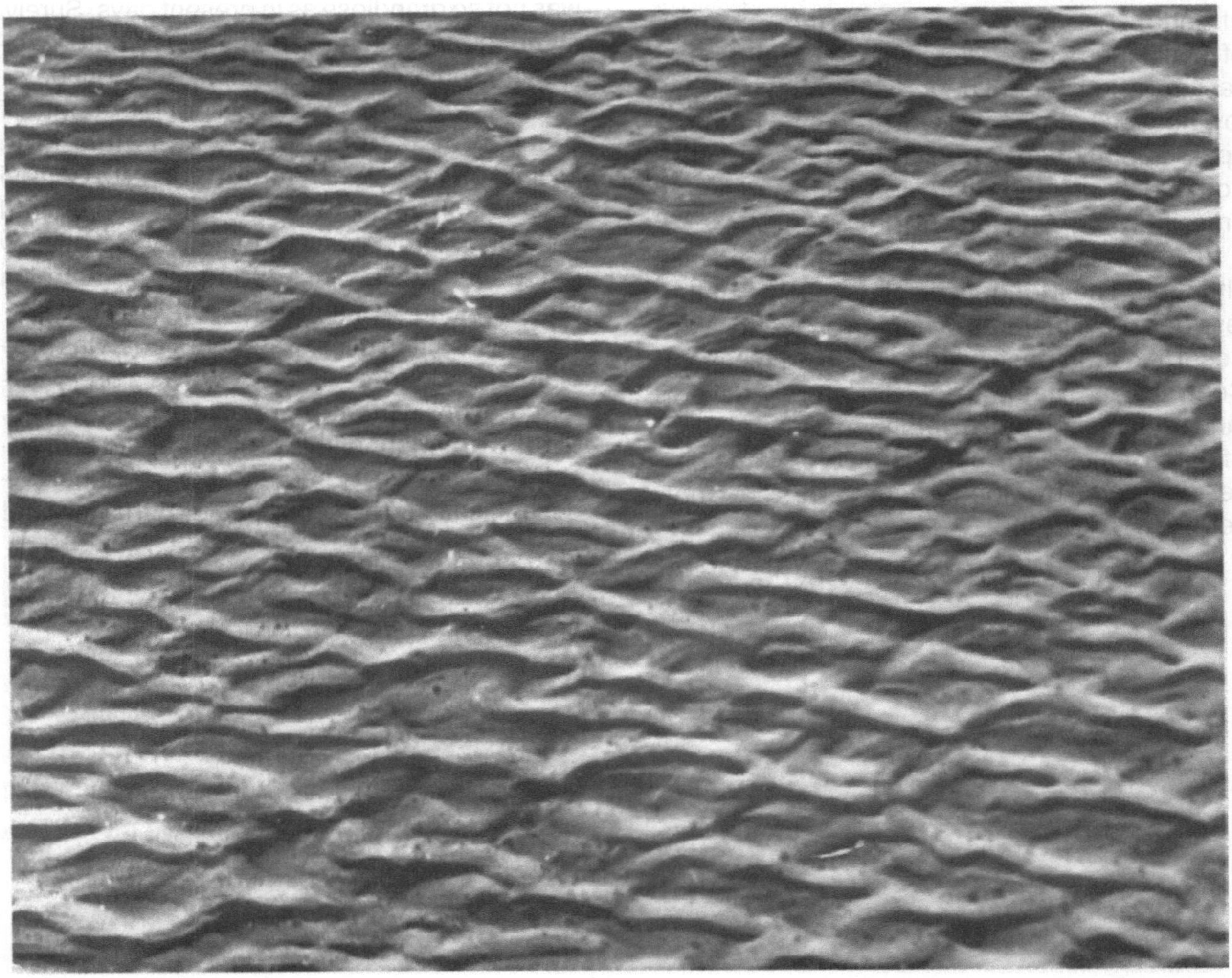

Figure 3.12. Near Amudarya sandy belt formed by severe human pressure. Aerial photo.

CHAPTER 4

POLITICAL, ECONOMIC AND SOCIAL IMPACT OF LAND DEGRADATION DURING THE SOVIET PERIOD

We shall come to the victory of the Communist labor.

V. I. Lenin

Remaking nature, development of new lands, re-education of Moslem society, corruption and "pripiska" were the main factors which led to ecological catastrophe in Central Asia.

4.1. Large leap – a Central Asian version

Communist leaders put task of re-education of people in the spirit of the communist morals. According to the communist doctrine, the human society passed in consecutive order through several stages of socioeconomic development: primitive society, slave-owning system, tribal system, feudalism, capitalism, socialism and communism. The last was considered as the final stage of development, as an "ideal society".

Before the revolution of 1917 the people of Central Asia remained at the stage of early feudalism or even lived in the tribal system. So there was a gap between their stage of development and socialism. This gap included one or two stages. But this contradiction didn't confuse communist leaders. They said, people could "jump over" one or even over two stages and enter socialism or communism. Khrushchev said: "The people of our generation will live in a communist society". Religion, morals and tribal traditions were neglected in this process. The national economy, according to this concept, was considered as "a basis of socialism".

Agrarian revolution in Central Asia

After the three years (1917 – 1920) of bloody war the Soviet power was established in Russian Turkestan. Many irrigation systems were damaged during this war. In 1918 Lenin signed a special decree on allocation of 150 mln. roubles for reconstruction of irrigation canals in Turkestan. In these works so called "folk-custom method of works" was widely applied. As known, this "method" is none other than a forced labor of great number of people. The rulers of the past, khans and beks also used this "method".

The 10th Congress of the RKP(b) passed in 1921 the resolution on a new economic policy (NEP). By estimation of a Soviet historian, "... the land reform has destroyed colonial survivals and has stimulated the political activities of working masses but the reform couldn't solve in full measure the problem of land and water use" (Atamamedov & Moshev, 1984).

put on trial without a special permission of raikom!... If a communist committed a crime, his personal affaire was investigated first of all by raikom. This inability of communists in policy and national economy was one of the causes of break-up of the Soviet empire. Selection of personnel in Central Asian Republics was based on principles of favoritism and ties of relationship. For example, in October 1989 the Central Committee of the Turkmenistan Communist Party dismissed from his duty B. Ataev, the first obkom secretary of Tashauz region. He was accused of "pripiska and distortion of Party's cadre policy". One third of the people in his administration were kinsmen from his aul (Loginov, 1989).

Collectivization in Central Asian republics was organized on the same scenario like in other regions of the USSR. According to Conquest (1986), 40 thousand families in Uzbekistan were dekulakized, representing 5% of farmers. Bais, the rich people of rural area, (Russian equivalent of kulak), were subjects of these operations. Collectivization in Central Asia gave impetus to organized resistance. Guerilla bands, known as basmach, appeared in many regions of Central Asia. In 1931 – 1934 some 350 guerilla bands fought in Uzbekistan. One hundred and sixty four cases of arm resistance to collectivization were registered in this republic. Thirteen thousand dekkhane participated in these clashes with governmental forces. In the Syrdarya valley, rebels resisted for two weeks. In Turkmenia many basmach bands were active untill 1936. In Kazakhstan 40,000 bai families were dekulakized, the rest left their flocks and scattered.

"Proletarian captain" S. Budennyi played an important role in neutralization of basmach bands. As Vladimirov (1993b) noted, Stalin said to Budennyi in March 1926: "Don't you see Semen Michailovich, that the campaign against basmach bands is continuing? The Central Committee has passed a resolution to send you on a mission to Central Asia. You have battle experience in fighting against guerilla bands". And he carried out his tasks. Mothers have been scaring children with the name of Budennyi until present days. Until last days of the USSR's existence, one couldn't see a portrait of Budennyi in state institutions of Central Asia...

Collectivization in Central Asia resulted first of all in reduction of livestock (Conquest, 1986). In 1929 livestock in Central Asia (including Kazakhstan) numbered 7,422 thousand head. By 1933 this number was reduced to 1,600 thousand head. The number of sheep was reduced from 21,934 thousand head to 1,727 thousand head during the same period. About 1.5 mln. people lost their life "in the process" of collectivization. In such a way the agrarian revolution in Central Asia was completed.

Rooting out the survivals of the past from the people's mind

The mode of life changed by collectivization. New settlements and administrative centers appeared in rural areas. Many new people came to auls, some of them were from other kin groups or even were aliens, for example Russians. It was a real cultural revolution conducted by instructions given "top-down". The struggle began against "bad feudal-patriarchal survivals and the dogma of Islam which held people in the captivity of obscurantism and medieval ignorance" (Durdyev, 1984). The slogan of Lenin "Religion is opiate of the masses" played very important role in this cultural revolution.

Party cells created in auls began the struggle against Moslem traditions and rites. New personages of a new political system appeared in rural area, in particular partorg, raikom secretary, head of kolkhoz (bashlyk) etc. Mullahs, well known leadership personages of Oriental tales, disappeared gradually from everyday life. The kolkhoz concentrated in its hands enormous power. Farmers were transformed to slaves deprived of the human rights.

Was it possible to combine the old traditions with a new mode of life? Vagabov (1968) gave an affirmative answer. He said, that some Moslems who had three wives, send them to work in the kolkhoz. But they themselves being the heads of families didn't work in the kolkhoz!..

The struggle for freedom of women was especially drastic. Uzbek and Tajik women covered their faces with a special veil – parandzha. Communists started a campaign against parandzha. As a result, 100,000 women in Uzbekistan took off parandzha in 1927. In the same year 4,000 women in this Republic learned to read and to write. But the losses of women in this struggle were great. Two hundred and twenty six women in Uzbekistan were killed.

Pig breeding could be also mentioned as one of the methods of struggle against "religious survivals in the people's mind". Before the revolution, pork was consumed in Central Asia by Russians and by other non-Moslem people. In 1916 the number of pigs in Central Asia totaled 30 thousand head, in 1941 it was 247 thousand head, by 1969 it reached 610 thousand heads (Central Asia, 1968). According to Sheripov (1984), pig breeding in the Turkmen SSR was a prosperous branch of the economy. That was especially clear, as he said, after the resolution approved by the KPSS Central Committee in March 1965. So the Moslem was forced to play the hypocrite... At present, many Moslems in Central Asia have been consuming pork, mainly boiled sausage.

Corruption in the Soviet bureaucratic system increased, the heads of kolkhoz and partorgs became hypocrites. They participated in party meetings, but secretly observed all Moslem rites. The hypocrisy penetrated also in the higher echelon of power. For example, B. Shif, a Bukhara Jew, was the second obkom Secretary in the Samarkand region. He belonged to the Khassids sect and secretly performed the duty of rabbi (Kondrashov, 1994).

Circumcision of boys, weddings and other festive occasions were celebrated during the epoch of "Building-up Socialism". Often these festivals were transformed into mass carousals with hard drinking and consumption of teriak.

Hard drinking and alcoholism became a real scourge of Moslem and non-Moslem population of Central Asia. After the well known decree of Gorbachev in 1985, the struggle against drinking began. Fisak (1988) published very interesting information on this subject. The town of Bairam-Ali in the Turkmen SSR was declared "the zone of soberness". But in 1987 police arrested in this zone 379 drunkards. During the same year 13.5 thousand drunkards were registered at "sobering"stations of this republic. The number of drunkards in this year increased: among members of the Communist Party by 1.7%, among members of Komsomol (Communist Youth Organization) by 8.8%, among officials by 14.8%. One could doubt that these figures were underestimated!..

Builders of the Karakum Canal also consumed vodka. In the Chardzhou region, an additional sum of 80,000 rouble was funded for purchase of vodka for this category of working people in 1987 (Khudainazarov, 1988). But the real figures of consumption of vodka were not published.

Party leaders couldn't abolish an old Moslem tradition – paying bride-money, so called "kalym"!.. To our great regret, this tradition was supported by women themselves. According to public opinion, marriage without kalym was shameful for woman. The size of kalym depended upon the standard of life. During the prerevolutionary time, kalym totaled 400 – 800 roubles. It was rather big sum of money taking into consideration the high value of tsarist roubles. Moser (1885) said that after the Geok-Tepe massacre, when many Turkmen were killed, prices of brides fell. In 1945 – 1955 kalym in the Turkmen SSR totaled 15 – 20 thousand roubles. That was equal to the cost of one cow. In the seventies kalym totaled 20 – 30 thousand roubles, that was equal to the cost of a car on the black market. In 1992 kalym totaled 200 – 400 thousand roubles because of inflation. These figures are based on confidential information from some people.

Marriages without kalym were so rare that these cases were usually discussed in the mass media. The women's council of Nebit-Dag town in the Turkmen SSR was very active in propaganda for a new way of life. The council was headed by D.Sakhatova, a progressive woman (Polovinkina, 1988b). This council approved the conduct of newly-married couples who rejected kalym. Some women according to this information, who earlier supported kalym during perestroika period, changed their mind and were transformed to supporters of the communist morals!.. Kh. Gel'dykhanova from the same town received 5,202 roubles as kalym for her daughter. This kalym was confiscated, according to the sentence of the law-court. Above that, the guilty paid a fine of 312 roubles (Anonymous, 1988a).

During perestroika, a campaign also was started against so called "unearned income". One of the first sociological studies was conducted in the Turkmen SSR. The results of this study were not published in full measure, only some of them became known (Il'yasov & Mukhametberdiev, 1988). 38.1% of people questioned gave an affirmetive answer on the question: "Did anybody demand a bribe from you?" 80.3% said that, by purchasing goods, smugglers rendered them service. 73% of bribed income came in state trade; the rest involved police, hospitals and chemists. Bribery income in science totaled 4.5%. The question also was asked if the administration took measures against unearned income or if it only pretended to do that. 45.4% answered that the administration only pretended to do that. More than a half of interviewed people said that police and justice institutions took bribes.

The problem of strengthening atheistic propaganda was discussed in the mass media up to1991. For example, Kuliev (1988) demanded that the struggle against Moslem religion in the Turkmen SSR be strengthened by coordination of activities of the governmental institutions. He demanded special funds be given to "The House of Atheistic Propaganda" in Ashkhabad which was not operating effectively.

Improvement of public health care

Improvement of medical service in Central Asia was one of the greatest achievements of the Soviet power. In 1919 – 1920 a campaign was conducted on eradication of spotted fever and relapsing fever (Kiyatkhanov & Tesler, 1984). The distribution of cholera and other infectious diseases was also eradicated. By 1941 great progress was achieved in eradication of malaria and dysentery. Sanitary conditions in rural area and in towns were improved. Especially great was the role of Russian doctors and medical assistants. The number of doctors in Central Asia is indicated in Table 4.1.

Table 4.1. Number of doctors in Russian Federation and in Central Asian republics per 10,000 people (National Economy of the USSR, 1991).

Republics	1940	1965	1970	1979	1987
Russian Federation	8.4	24.3	27.5	35.4	46.3
Kazakh SSR	4.3	18.7	21.8	30.8	38.0
Kirghiz SSR	3.8	19.1	20.8	27.8	35.1
Uzbek SSR	4.7	17.0	20.0	27.7	34.7
Tajik SSR	4.1	15.0	15.9	23.1	27.2
Turkmen SSR	7.6	21.2	21.4	27.3	34.4

Table 4.1 testifies to the fact that the level of medical service in Central Asia was lower than in Russian Federation. But really it was even lower because the qualification of local doctors was lower than in Russia. Local doctors recognized that (Mamedov, 1988). Young people joining medical institutions in Central Asia didn't master the Russian language, so they had difficulties in the first years of training.

The number of beds in hospitals is an other criterion of medical service. Table 4.2 gives this information.

Table 4.2. Number of beds in hospitals of the Russian Federation and in Central Asian republics, per 10,000 people (National Economy of the USSR, 1991)

Republics	1940	1960	1970	1980	1987
Russian Federation	43.3	82.1	112.5	129.6	135.4
Kazakh SSR	39.5	80.4	118.4	130.1	135.1
Kirghiz SSR	24.1	73.5	106.7	119.7	118.6
Uzbek SSR	30.1	83.9	101.7	113.1	120.4
Tajik SSR	28.6	67.1	99.7	98.8	104.2
Turkmen SSR	41.6	83.0	101.4	104.5	109.2

In conclusion of this section we want to discuss one point. Owing to what, or precisely owing to whom were these results achieved? The only answer is the following: on the account of Slavic peoples, mainly Russians. The central government in Moscow stimulated the development of Central Asian nations by conducting a special policy prejudicial Slavic nations. During the Soviet period, agriculture in Central Asia was 10–15 times more profitable than in Russia. Prokhvatilov (1992) explained this as clue to ignoring economic laws. According to the economic theory of socialism, land rent was effective only in the capitalist society, not under socialism. So, the land rent was not taken into account by planning the development of agriculture. As a result the Slavic nations became poorer and the people of Central Asia became richer.

Moscow funded enormous sums of money for economic development and medical service in Central Asia. According to Durdyev (1984), in 1932 the expenses on public education totaled (roubles per capita): in Russia – 2.77; in Ukraine – 5.92; in Tajikistan, 18.8; in Turkmenistan – 23.79. The number of schools in the USSR increased during 1914 – 1941 by 1.8 times , that in Turkmenistan by 21.6 times. This policy led to the reduction of population in Russia. By the beginning of perestroika, thousands of Russian villages were abandoned. Farmers *en mass* left the rural areas and tried to find job in towns and in industrial centers. During this time many new houses were built in auls of Central Asia and a rural population had a higher standard of living to be compared with the people in Russian villages.

4.2 Cadre determines all!

"Cadre determines all!" This slogan was a favorite of Stalin. The slogan was especially useful in Central Asia, where 98% of population were illiterate before the Bol'shevik revolution. A campaign for universal literacy began in 1929. Numerous schools were opened in auls and towns. This literacy campaign was very difficult and long, especially in training adults. But the progress in education was remarkable. For example, 82% of population in Turkmen SSR could read and write by 1941 (Durdyev, 1984).

Training cadre was done on a "conveyor system". Instructions on education were given top-down, and the criteria of education were quantitative, but not qualitative. Party leaders of Central Asian republics, obkom and raikom secretaries estimated the quota of specialists of local nations to be trained. That resulted in graduation of a great number of natives whose level of education was very low, lower than that of Russians.

Training personnel was first of all considered from the ideological point of view. According to the ideological concept of communists, a physician who didn't study

Marx – Lenin theory could not treat people for an illness... Communist philosophy became the principle subject of studying in universities. On the account of that, the length of university education became longer by 1 – 2 years.

As known, the system of education in all countries consists of three stages: pre-school, school and university education. Traditionally, Moslem families were always large. Parents being engaged in heavy labor on cotton plantations couldn't take much care of their children. As a result, mental development of small children was low. In several cases this backwardness couldn't be compensated for in the next stages of education. As said: " The train has departed" ... In auls of Central Asia many debile (mentally deficient) children were registered. Among them were children with inborn defects and children acquired deficiencies. The first category was mainly consisted of children whose parents were close relatives. Among Moslems of Central Asia, marriages were often concluded between first cousins. In this case kalym was very low, often it was symbolic. Poor people sometimes behaved in such a way if they didn't have money for kalym.

Mamedov (1988) found that acquired deficiencies were found in children who didn't go to kindergardens. Only 13% of the children in auls of the Turkmen SSR were brought up in kinder-gardens; that was much lower to be compared with towns. So the children in auls couldn't read before entering schools. In towns many children could do that.

The 23rd Congress of the KPSS passed in 1966 passed a special decree on compulsory secondary education. All children, in spite of their natural gifts or their wishes has to get school-leaving certificates. S. Niyazov (1988), the former First Secretary of the Communist Party of Turkmenistan, criticized this system. He said that in Turkmenistan each third or even each second pupil leaving the secondary school did't make adequate progress in training. In spite of that, all they received school-leaving certificates!.. That was a real "pripiska" in education. In 1988 the shortage of teachers in Turkmenistan totaled: in mathematics – 945; in Russian language – 885; in Turkmen language –100; in labor training – 250.

Slutnik (1988) described a sad state of people's education in Tashauz region of the Turkmen SSR. Shortages of school buildings and poor sanitary conditions were common in many auls and settlements. 290 public schools occupied saman houses, 100 of them needed emergency repair. Teaching in 60% of the schools was conducted in three shifts. Schools of the region needed 800 extra teachers in the Russian language.

Teaching in national schols (in which children of native nations dominated) was conducted on a low level. Suleimanova (1988) told about the low level of mental development of children in public schools of the Turkmen SSR. In one school, children had to prepare a composition on the subject " The white gold". According to Suleimanova's opinion, this subject was close to their interests because they worked on cotton plantations. But childred didn't fulfil this task. In their opinion it was hard and needless work.

Teaching in natural history, botany and biology was conducted formally. After these lessons, pupils broke young trees planted around schools by their parents. School discipline was very low. According to Ryabova (1988), in one school of Ashkhabad a school mistress hurled a bucket at a pupil who did not obey school order. The same school mistress tried to get sausage and butter from those pupils whose parents were engaged in commerce. It was "a duty paid in kind"... To our great regret, the name of this school mistress was unknown.

Participation of school children and teachers in so called "subbotnik" was obligatory. Subbotnik was usually confined to the birth day of Lenin. Children of Ashkhabad planted trees and cleaned streets on this day. According to the official point of view, subbotnik was the best method of ecological education of children. But Niyazov (1988), being the First Secretary of the Communist Party of Turkmenistan, criticized subbotnik conducted in Turkmenistan. Children cleaned the streets and planted trees, but they didn't receive money for this job. The money was paid to the municipal workers who were responsible for this job...

Children's labor was widely used in agriculture, especially by cultivation of cotton. Bairamov (1988) published the following information on this subject. The following mean number of days per year was worked in kolkhoz of the Turkmen SSR:

	1975	1980	1985
Men	253	252	260
Women	191	202	206
Youth*	119	124	137

*Younger than 15 year.

According to labor regulations, the youths could work only 4 hours per day, but in fact they worked the whole day (Mamedov, 1988).Children of secondary schools in rural areas, belonging to local groups, couldn't master the school programs because of the work on cotton plantations. After 10 years of training, the main part of them could only read, write and do simple sums.

In spite of that all of them received school- leaving sertificates. An addition, they couldn't speak Russian fluently. Some of them joined universities. During the first years of Soviet power, young people of local groups entered universities without entrance exams. Later this system was changed. Candidates had to pass exams but local people had privileges. A corrupted system of training cadre was fully formed by the beginning of the seventies. Charukhchev (1988) described the panic which embraced the parents of young people in Ashkhabad who wanted to join universities. What was the size of baksheesh, to whom should it be paid etc...

All students of universities had to work for 2 – 3 months on cotton plantations. This work was considered as "practical training". During this forced labor, students lived without any comfort; their food was of bad quality, they couldn't take baths and medical service was not provided. As known, experienced cotton-pickers could collect about 100 kg of row cotton per day. The main productivity of labour in the kolkhoz of Turkmen SSR was 35 kg of row cotton per day (Osipov,

1988). Students could collect only 30% of this rate, because they were not interested in this job. So, from an economic point of view, their labor on cotton plantations was unreasonable. This job was only very important from the ideological considerations.

Drinking alcohol was the main amusement of students during the campaign of cotton-picking. Often scuffles arose between students of different clans or different insitutions. As informed by Kovalevsky (1985), students in one kolkhoz of the Turkmen SSR arranged a scuffle. Their instructer, M. Annamukhamedov, lecturer of the Turkmen Polytechnical Institute, got into "sobering" station... Student work on cotton plantations was banned only during perestroika.

Even more grotesque was granting degree by correspondence. Many native administrators wanted to have university diplomas for advancement in their careers. For example, the director of one of the industrial enterprises in Ashkhabad was a student by correspondence in the Turkmen Polytechnical Institute. He paid lecturers of this institut for projects which they didn't implemented and for passing exams. Money for this affair was taken from public coffers. The author of this information (Kovalevsky, 1985) didn't give the names of the participants in this affair.

Ecological education of population was organized through many so called public organizations: "Society Knowledge", Society of Nature Conservation, Geographical Society, Botanical Society, etc. But stricktly speaking, all these public organizations were not non-governmnetal organizations because they stayed under control of party personal. They mainly fulfiled ideological functions. Formalism and pripiska were characteristic features in activities of these organizations. For example,Mukhammedov (1988) published very interesting information on activities of "Society Knowledge" in the Turkmen SSR. In Geok Tepe district, "Society Knowledge" had many "active lecturers", numbering 350. They delivered 2,000 lectures for the local population per year. According to information of the central office of this society in Ashkhabad, 7.2 mln. people attended popular lectures in 1987 (population of the Republic being 4 mln.!..).

Public education and ecological training should provide for the development of ecological concepts in the people's mind. The local cadre determined ecological policy during Soviet period. The three main points should be emphsized here:

1. In spite of all shortcomings, education of people and training the local cadre was one of the greatest achievements of the cultural revolution in Central Asia.
2. Low level of ecological thinking of the local cadre was one of the negative factors in nature conservation.
3. Eradication of religious feeling from the consciousness of young people led to their moral degradation and that in its turn stimulated its absolute obidience to party policy.

4.3. Change of social and ecological conditions in rural area

Agriculture in Central Asia during the pre-revolution period was a pure branch of the economy (from the ecological point of view). Soviet leaders demanded the increase of cotton production, this so called "white gold". Every year the communist administration in Central Asian republics called on the people to work on cotton plantations, especially in cotton picking.

The task could be solved by development of new land, application of machinery and chemicals and by participation of all population in agricultural works. Dyuzhev (1984) gave the figures on the application of chemicals in the Turkmen SSR. Weeding of cotton planations was conducted by application of chemicals (herbicides). In 1970 herbicides were used on 14.9 thousand hectares, in 1975 on 97.4 thousand hectares, and in 1982 on 223.7 thousand hectares. Chemicals were also used by mechanized cotton picking: defoliants for removal of cotton leaves and dessicants for drying up the plants. This treatment was conducted in 1970 on 452.9

thousand ha, in 1974 on 559.2 thousand ha and 1981 on 913.8 thousand ha. The author of this publication emphasized the importance of application of chemicals in agriculture but he ignored the ecological problems.

The town of Mary in the Turkmen SSR suffered from a shortage of potable water. After the construction of the Karakum Canal, its water was conducted to the city water supply system without purification. And only 25 years later a project was developed on construction of a purification station.

The work on cotton plantations was obligatory for all people, from 8 years and up. Only aged people who couldn't move were released from this job. The people who didn't work were punished administratively or by application of economic sanctions. For example, the administration of a kolkhoz could cut off their personal plots of land. For poor people that was equal to death from hunger.

During the period of perestroika many cases of obuse of power were revealed by the mass media. For example, a certain A. Adylov, the head of the agricultural association in the Uzbek SSR treated collective farmers like slaves. He punished lazy slaves with a lash or put them to gindan. This gindan, specially constructed for this purpose, was shown many times in TV program from Moscow.

The case of Adylov is of great interest. He was arrested the first time in 1981, being charged with pripiska, bribes and assassination of a raikom secretary. A special comission of the Supreme Soviet of the Uzbek SSR investigated this case and Adylov was acquitted of the crime. In 1984 he was arrested again (Vasil'ev, 1991). The Supreme Court of the USSR investigated this case during 6 years. Adylov was charged with plundering 1 mln. roubles, taking bribes in the sum of 600 thousand roubles, etc. The author of this publication said: " Adylov would be sentenced soon". But he was never sentenced. After the desintegration of the USSR, criminals in Uzbekistan became "the victims of the Soviet regime". He was again shown by Moscow TV in 1992. He was quite well, working on his personal plot of land...

In the Turkmen SSR, many tracts of newly developed land were located at the distance of 70 – 150 km from villages where cotton growers lived. Temporary huts were built for them when they worked on cotton plantations, often with their wives and children. They lived in these huts in unsanitary conditions. They used dirty water from canals for cooking and drinking (Charykliev, 1988).

New settlements were built in the cotton cultivation areas of the same republic. For example, a new village was set up in Khauzkhan tract , but for 10 years neither baths nor water lines were constructed. The settlement had only one telephone, the medical post was poorly equipped, only drugs for headaches were available. A club and library occupied a small space but nobody visited them (Karpov, 1988).

In the same Khauzkhan tract, labor protection was fully ignored. Kolyabin (1988) told about participation of young girls in the chemical treatment of cotton seeds before sowing. A strongly harmful chemical substance was used for this purpose. But the girls were provided neither with protective gloves nor masks.

As a rule, some males worked on cotton plantations of Central Asia, being team leaders, drivers, tractor operators, etc. Many of them preferred to be engaged in commerce or to pass time in chaikhana. All hard and unskilled labour was fulfilled by women and children (Myalo & Goncharov, 1989). According to Borodavka (1988), in a kolkhoz named after Lenin (what a name!) in the Turkmen SSR 60% of the collective farmers didn't work in the kolkhoz. They worked on their own plots of land or were engaged in commerce in the black market.

A special commission of the Congress of Deputies of the USSR studied the life of cotton growers in Uzbekistan (Kovalev, 1990). Living conditions of people in villages were terrible. There were many families having 10 and more children. In their small houses the members of the Commission could see only articles of bedding. The heads of families earned 100 –150 roubles per month; that was below the a subsistence minimum...

The avarage size of families in Central Asian republics was as follows: Kazakhstan – 4.0;

Kirghizstan– 4.7; Tajikistan – 6.1; Turkmenistan 5.6; Uzbekistan –5.5 (National Economy of the USSR, 1991). Infant mortality in Central Asian publics was higher than in Russia (Table 4.3).

Table 4.3. Number of infants who died before reaching 1 year, per 1,000 born (National Economy of the USSR, 1991).

Republics	1970	1980	1985	1987
Russia	33.0	22.1	20.7	19.4
Kazakhstan				
Kirghizstan	45.4	43.2	41.9	37.8
Tajikistan	45.9	58.1	46.6	48.9
Turkmenistan	46.1	53.6	52.4	56.4
Uzbekistan	31.0	47.0	45.3	45.9

But we may suppose that the real figures of infant mortality were higher because of the Moslem tradition. According to this tradition, registration of birth and death is not obligatory in Moslem communities. So in several cases people in rural areas buried infants without any registration of two facts: the birth and the death.

The status of Moslem women was terrible. Only 1/3 of women in the Turkmen SSR who had children younger than 1.5 years received a state allowance for large families (Bairamov, 1988). The rest lost their right because the interval between births was too short. To receive this allowance, women had to work before the childbirth. Moreover, many women ignored this allowance because it was very small.

The low cultural level of Moslem women was also one of the reasons for high infant mortality. For example, many Turkmen women ignored inoculation of small children. Instead of that, they tied a special colored thread (aladzha) around the child's wrist. This aladzha in their opinion protected children from magic and infection (Mukhammetberdiev, 1988).

Agricultural products in Central Asia contained nitrates, nitrites and pesticides in inadmissible quantities. Fruits and vegetables were sold on markets without any control (Seiitkurbanov, 1991). By a rough estimation, 50% of vegetables were lost by transportation and storing. In the Turkmen SSR only 40% of works in storing places were mechanized, only 28% of store houses had air conditioning and only 15% had active ventilation (Bairamov, 1988).

Use of dope was always very popular among natives in Central Asia. But the number of drug was secret information which never was published during the Soviet period. The first information on this problem was published after 1985. Tachmamedov (1989) said that the administration of Turkmenistan undertook insufficient efforts in the struggle against use of dope. In 1989 local police discovered 162 sites where opium poppy was grown. In these places 245 thousand poppy plants were growing. The campaign against use of narcotics was continued in the Turkmen SSR in 1989 (Anonymous, 1989). Many places of cultivation of opium poppy were discovered. For example, a collective farmer from Iolotan district, G. Rakhmedov had grown in his personal plot 514 plants of opium poppy.

4.4. "Pripiska" and its role in land degradation

Mass falsification of information, or "pripiska".

What is "pripiska"? It can be defined as "*intentional falsification of accounts and reports aimed at obtaining illegal income*". But this definition characterizes only a lawful aspect of the problem. During the Soviet period, pripiska became a mighty factor in political development, in public life and in ecology. It was also a serious factor of land degradation.

We may say *a priori* that pripiska appeared simultineously with written languages, accounts and reports. Many historical sources have confirmed that. For example, Mez (1922) said that pripiska had been a standard practice of administration in the Baghdad Khaliphate during the Middle Ages. He quoted an Arab historian who said: "The head of the department got money to pay the salary of persons who didn't work in his office or who didn't exist at all. It was done by falsification of the lists of salaried workers, in particular clerks and lawyers. Expenditures for paper were also overestimated and a part of the paper was sold aside". These events occured in the 12th century A.D. But one may imagine, this historian lived in the 20th century A.D.!

Pripiska was also common in Central Asia during the pre-revolutionaty period. Lieutenant-colonel Grulev (1901) told about falsification of reports in the Bukhara Khanate. Bek (the governor of the province) was responsible for collection of taxes which he had to deliver to the Emir's treasury in Bukhara. Bek had two books for registration of taxes, one of them contained the real figures, and the second book had false information. This last book could be shown in case of inspection.

Pripiska was also common in the Semirechie region (Palen, 1910a). The Kirghizes always gave lower figures about livestock numbers and about areas suitable for cultivation. They did that because of two reasons:

1. According to traditional belief, people must never give true information about their property because it could do harm,
2. They were afraid that Russian migrants would occupy the land suitable for cultivation.

During the Soviet period, pripiska became a mighty social factor which transformed people's morals. The size of pripiska in agriculture, industry and science was catastrophic. Pripiska, corruption and moral decay of many people were especially great during the so called "Period of Stagnation" (zastoi) which was identified by the communists themselves with the rule of Brezhnev. N.Voronov, the former member of the Politburo said that Brezhnev didn't response to signals about corruption and pripiska in Uzbekistan and Azerbaijan (Lynev, 1988). The following main features of Soviet pripiska should be emphasized:

1. Pripiska became a mass event, that was not common during the pre-revolutionary period,
2. Pripiska became almost a legal deed which was not condemned by society,
3. Pripiska became a part of the image of a Soviet administrator, as a measure of his maturity and his ability to rule over people and to be on good terms with the authority.

In what way did pripiska concern the problem of land degradation? A cotton affair in Uzbekistan can clarify this issue.

Cotton affair in Uzbekistan

Many Uzbeks have been feeling hurt about the event called the "Uzbek cotton affair". On this point we can agree with them because the cotton affair and other forms of corruption were common in other Central Asian republics. But in Uzbekistan, pripiska and corruption descovered during the last years of the existence of the USSR were so terrible that the "Uzbekistan case" was a "classical" case which would be remembered for a long time in the annals of history.

A Russian proverb says: " Fish goes bad from the head". Moral degradation of the society began from the head, from the ruling clique (Nishanov, 1988a; Sakhnin, 1988). For example, R. Kurbanov, the Chairman of the Council of Ministers of the Uzbek SSR was sentenced to 8 years of imprisonment because of bribes and corruption. Myalo & Goncharov (1989) called the political system of the Uzbek SSR as a "satrapy of the 20th century". Corruption of the ruling clique, a shady economy and traditions of clans were joined in this system.

The monoculture of cotton transformed the people of Uzbekistan to slaves. The system functioned on a German proverb: "Der Ober drückt den Unter". That means "the senior oppresses the junior". The first Central Committee secretary squised the obkom secretary, the obkom secretary squised the raikom secretary, the last squised the head of the kolkhoz, etc. The plain people (dekkhane) who represented the base of this pyramid carried its weight...

Pripiska in production of cotton was common on all levels. First of all, the size of sown areas was underestimated in statistical reports. New land from reserves was used for cultivation of cotton. The reports on mean cotton yield per hectare were falsified.

Another type of pripiska was occurred in storage places of cotton, where the quality of the cotton was falsified. But the most serious pripiska included the registration of cotton which was not produced. Moscow demanded each year to increase the production of "the white gold". Local party functionaries offered a counter plan which exceeded the quota given from Moscow. Ultimately, all reserve lands for cotton production were exhausted, but Moscow continued "to press the button", as the local people said. Moscow demanded increased thousands of metric tons of cotton... Nishanov (1988 b), the former First Secretary of the Communist Party of the Uzbekistan said: " Bribes and corruption in agriculture and in industry, especially in cotton production during the last decades, led to severe consequences". Among these consequences he mentioned drying up the Aral Sea.

Correspondent Zhbanov (1990) described the beginning of the investigation of the cotton affair, when two carriages loaded with lint (waste product of cotton) were transported from Uzbekistan to Serpukhov (Russia), where they were arrested. Inspector K. Maidanyuk discovered that mln. of metric tons of cotton were registered in Uzbekistan by false documents during the rule of Sh. Rashidov and 2 billion roubles were plundered. The Supreme Court of the Uzbek SSR condemned to death V. Usmanov, the Minister of Cotton Industry. The members of the higher board of this ministry were also sentenced. Later U. Usmankhodzhaev, the First Secretary of the Central Committee of the Communista Party of Uzbekistanan was also sentenced for bribery and corruption. Inspector Maidanyuk resigned after that under the pressure of the mafia...

Inspectors T. Gdlyan and N. Ivanov continued the investigation of the cotton affair. They reported many times in TV programs from Moscow about the results of this process. They both became real "film stars" of perestroika. Later, in 1993 Inspector Ivanov was murdered under mysterious circumstances (Luk'yanenko, 1993).

Pripiska was also common in other regions of Uzbekistan. Sadykov (1988), the former obkom Secretary of the Karakalpak Autonomous Republic, said that all reports on achievements in agricultural production in Uzbekistan during the 12th Five-Year Plan Period were false. During the last years, production of milk per milking cow in Kararalpakia was falsified on 1,000 kg per year per cow. Production of corn was falsified two times. About 1/4 of the sown areas of cotton and rice were not registered.

Pripiska and corruption in other republics of Central Asia

Pripiska was also a common practice in other Republics of Central Asia. Several examples are given below. Charykliev (1988) described the case of pripiska in the Turkmen SSR. Two collective farms "Kyzyl Baidagy" and "Kuibyshev" in the Murgab administrative district cultivated cotton illeagaly on the land that belonged to Serakhs district. They tilled 2,000 hectares of land belonging to Serakhs district, but this area was not registered in reports. Row cotton collected here was purchased by the governmental agencies as produced by the two kolkhozes mentioned above.

A certain R. Durdyev, the book-keeper of kolkhoz "15 Years of October Revolution" in the same republic built a big house for his family which looked like a palace. He had four servants, including a personal cook and a barber (Kuleshov, 1988). But the account of the kolkhoz in the state bank was closed because this kolkhoz was bankrupt. A. Tedzhenova, a team leader of the kolkhoz, and some other people tried to make complaints against him but in vain. A. Bekmuradov, the public prosecutor of the district, protected the embezzler of public property.

Corruption in administrative and party bodies of the Turkmen SSR came to light during perestroika. According to Kuleshov (1990b), in a party body of the Chardzhou region mafia created several bands which murdered D. Rozybekova, the insrtuctor of obkom, in december 1987. In April 1988 the culprits were centenced.

Plundering and corruption in processing agricultural products were registered in the Turkmen SSR. S. Niyazov (1988) reported that "losses" of meat products in slaughtering houses of Ashkhabad, Chardzhou and Mary totaled 10 – 14%. Frankly speaking, the workers of these enterprises pilfered meat...

Was it possible to eradicate these criminal deeds? Inspectors of the Central Commettee of the Communist Party of Turkmenistan investigated corruption and pripiska in the Tashauz region. As a result 1,380 party functionaries were dismissed from their duty and were reprimanded. Thirty one members of the Central

Committee were removed from this body. Eighty deputies of the Supreme Soviet were deprived of their mandates (Niyazov, 1988).

4.5. Conclusion

Ideological and political factors played an important role in land reclamation in Central Asia during the Soviet period. But political leaders and schemers had a low level of ecological thinking and they didn't know what environmental changes would occur after realization of the projects on land reclamation (Figure 4.1).

Based on a planned economy the quotas on production of cotton, karakul and other products of agriculture were given by the central government from Moscow. These planned figures demanded the increased production of agricultural products from year to year.

The Soviet political system transformed the peoples of Central Asia to slaves. Men, women, children and old people worked on cotton plantation and received a miserable salary. Forced labour on cotton plantations was obligatory for various strata of society, especially for students.

The morale of people (from high Communist functionaries to plain cotton-pickers!) degraded. Corruption and pripiska propagated among all classes of society.

By rough estimation, pripiska in agriculture and livestock raising affected 20 – 25% of the production. Pripiska led to the destruction of ecological concepts in the minds of plain people and undermined trust in the administration.

Pripiska played an important role in the process of land degradation because all actors of this criminal deed neglected measures on land conservation.

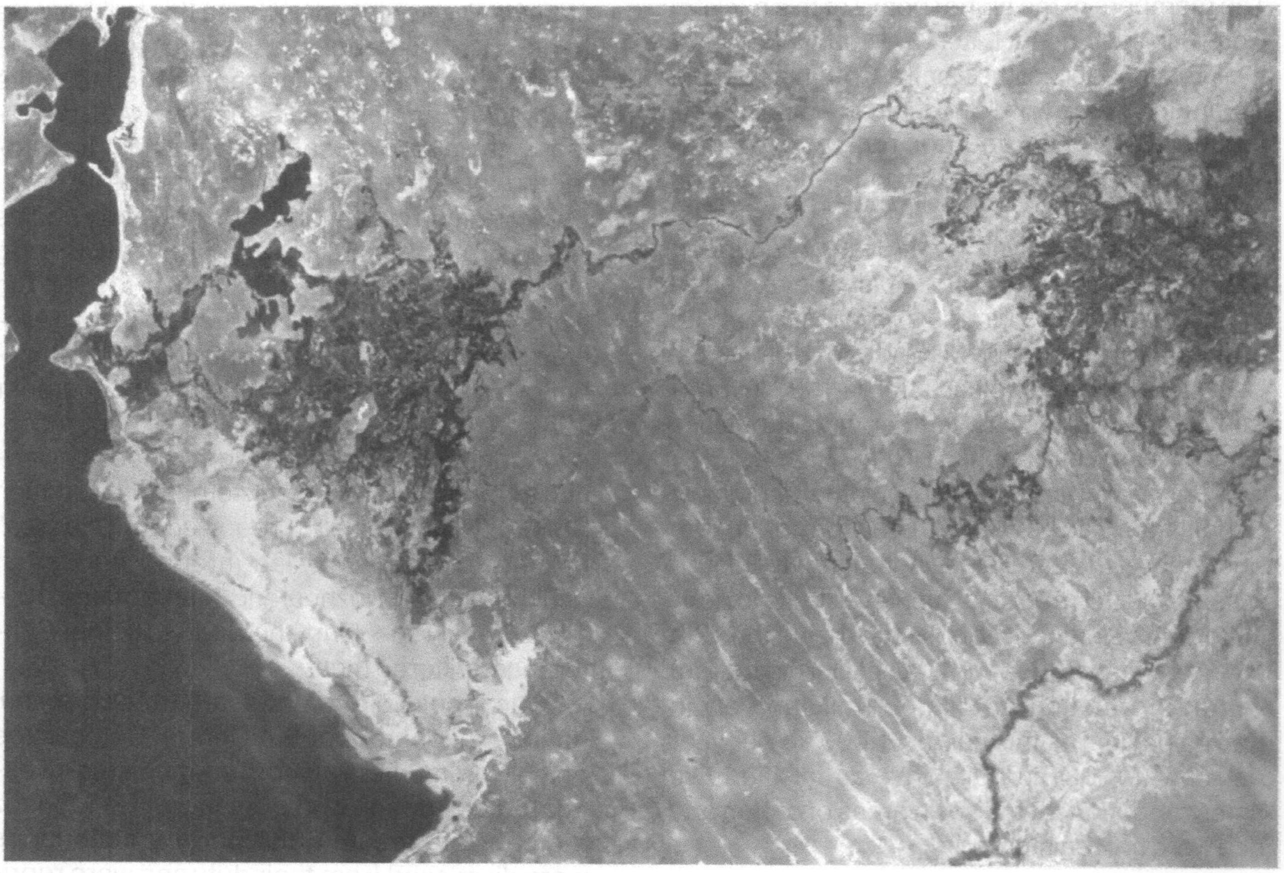

Figure 4.1. Space photo covering the delta of the Syrdarya and the costal strip of the Aral Sea.

CHAPTER 5

METHODOLOGY OF DESERTIFICATION ASSESSMENT AND MAPPING

Man can effect pleasant changes on the face of the Earth, but always at the risk of creating ugliness.
W. R. Rickmers – The Duab of Turkestan.

5.1. The concept and criteria of assessment

Conceptual approach for desertification assessment

According to the United Nations Convention to Combat Desertification (CCD), desertification means land degradation in arid, semi-arid and dry sub-humid areas. CCD explains, the term "land" means the terrestrial bio-productive system that comprises soil, vegetation, other biota, and the ecological and hydrological processes that operate within the system. So the land is the object of assessment by desertification study, not soil, not vegetation, but the land as an object of human activities. Degradation of soils or vegetation, degradation of arid ecosystems, degradation of landscapes are not synonyms, because these terms relate to other processes which also can be studied.

In our study conducted in Central Asia the following classification of lands used in the national economy was applied:

- Forests
- Desert rangelands
- Pastures
- Hay making lands
- Lands of dry agriculture
- Irrigated farmlands
- Gardens and vineyards.

Several land use (land cover) classes were excluded from assessment. To them belong:

- Swamps
- Moving sand dunes of the natural origin
- Stone and rock surfaces.

Human activities on these land cover classes is very limited, and their status is not caused by the interference of people. For example, a tract of moving sand dunes in Turkmenistan known as Dzhillikum (lit. "Furious Sands") was excluded from assessment because their origin is natural. Human settlements in the desert were also excluded from assessment, but not surrounding areas where desertification can occur.

The CCD notes the main processes of land degradation, such as:

- Soil erosion caused by wind and water
- Deterioration of the physical, chemical and biological or economic properties of soil,
- Long-term loss of natural vegetation.

In our project we have identified the following types of land degradation (desertification):

- Degradation of the vegetative cover
- Wind erosion
- Water erosion
- Soil salinization caused by the drop of the Aral Sea level
- Waterlogging of rangelands
- Technogenic desertification.
- Salinization of irrigated farmlands.

The number of desertification types depends on local conditions, methodology of assessment and the scale of the map to be prepared. For example, the scientists of India (Singh, Kar & Joshi et al.,1992) identified the following main desertification types: wind erosion and sand accumulation, water erosion, sodic soil salinization and formation of bogs on irrigated farmlands. Above that, they identified natural processes, for example wind erosion and sand accumulation. Methodology of FAO/ UNEP (Provisional methodology..,1983) indicates some other desertification processes, namely: soil crusting and compaction, reduction of organic matter, and excess of toxic substances.

In publications of Russian geographers the terms "wind erosion" and "deflation" are used as synonyms. In some case the term "deflation" is used for characteristic of natural processes, and the term "wind erosion" characterizes anthropogenic processes. In geographical glossary (Stamp, 1961) the term "deflation" is defined as the process of blowing out sand particles and their transportation by the wind.

Water erosion, in the wide meaning of the word, is lowering the land surface by the action of rain and water flows. But we must distinguish natural and anthropogenic factors. Gicheru (1996) indicated the main factors which cause the water erosion, namely: soil cover, resistance of land to erosion, slopes, characteristics of surface run off and the type of erosion (splash, gully, rill, etc.). By a practical study of water erosion, we can often see all these types of water erosion or their combinations within one land unit.

Each of these types of land degradation relates to different land use types.

Methodology of FAO/UNEP (Provisional methodology.., 1981) identified so-called aspects of desertification:

- Current status (CS)
- Current rate (CR)
- Inherent risk (IR)
- Human pressure (HP)
- Animal pressure (AP)
- Desertification hazards (DH).

CS is the change that occurred during a given period in a given region. CR is fixed ratio between two different status of desertification. The methodology of FAO/UNEP (Provisional methodology.., 1983) defines DH as "a source of danger of desertification, expression of effect of status of desertification, rate of desertification, inherent risk of desertification and human and animal pressure on environment". The methodology of FAO/ UNEP gives scales for assessment of HP and AP, but in our study we have selected the local criteria, suitable for Central Asia.

Scales for desertification assessment

By development of the criteria of assessment we have taken into consideration the world experience and our personal approach based on investigations in Central Asia, Mali and Mongolia. Criteria of assessment can be descriptive or quantitative. By compilation of the world map of soil degradation (Oldeman, Hakkeling & Sombroek ,1991) the following four levels of assessment were used:

1. **Light**: The terrain has somewhat reduced agricultural suitability, but is suitable for use in local farming systems. Restoration to full productivity is possible by modification of the management system at farmer's level. Original biotic functions are still largely intact.

2. **Moderate**: The terrain has greatly reduced agricultural productivity but is still suitable for use in local farming system. Major improvements are required to restore productivity partly at farmer's level, partly with government subsidies. Original biotic functions are partially destroyed.

3. **Strong**: The terrain is non reclaimable at farm level. Major engineering works are required for terrain restoration. Original biotic functions are largely destroyed.

4. **Extreme**: The terrain is unreclaimable and beyond restoration. Original biotic functions are fully destroyed.

As the authors of this methodology explained, these criteria give information about the changes of agricultural suitability, in relation to declined productivity and in relation to the biotic factors. In our study the criteria of assessment were developed for compilation of desertification map, scale 1:2,500,000. The object of our assessment was different from that selected by GLASOD methodology, as was mentioned above. The scale of assessment included the following three classes:

Level of land degradation	Classes of assessment
1. From unchanged to slightly changed	Slight
2. Moderately changed	Moderate
3. From severely changed to the total loss of the biological productivity	Severe

To estimate desertification, we must know its zero (a background) level corresponding to its natural (not disturbed) level. But in Central Asia, as well as in most arid regions of the world, there is practically no land undisturbed by man. The scales for desertification assessment are the following:

1. Scale consisting of five classes. Using this scale, one has no knowledge of the zero level of desertification, instead a given conditions of land degradation in a certain time period can be used. For example, by preparation of the desertification map of the USSR arid lands (scale 1: 2,500,000) we used 1965 desertification status as a zero level of assessment, than evaluated the changes that occurred from 1965 till 1988 (Kharin, Kalenov & Kiril'trseva et al.,1988).

2. Scale consisting of four classes. This approach can be used in two cases:

a) when a zero level is known or can be calculated, for example on long protected territories (natural reserves),

b) when a zero level is unknown, but the conceptual approach recommends to include into slightly desertified lands the territories "from not disturbed to slightly disturbed" and when the number of assessment classes is four. This approach is usually used by large and medium scales of mapping.

3. **Three classes scale**. A zero level is unknown and the scale of the map is small. This approach was used by preparation of a small scale desertification map of Turkmenistan (See Section 5.3 of this book).

Various types of processes characterizing desertification can be used as indicators of desertification. Diagnostics and monitoring of desertification are based upon criteria which allow quantitative and qualitative assessment of these processes. Local desertification criteria characterize this process within separate land cover by types. They can include, for example data on species composition and vegetation density in specific plant communities. We have used such criteria for preparation of large scale desertification maps. Regional criteria are usually developed for geographical regions or countries, for example for Turkmenistan. Zonal criteria of desertification allow study of vast regions, for example the arid zone of Central Asia.

The assessment of desertification is complicated by the presence of various factors. For example, in the Karakum desert the decrease of area covered by vegetation and development of wind erosion are caused by shrub logging for fuel, by overgrazing and the movement of motor vehicles. These factors can be assessed only if the map scale is large whereas small scale maps should either reflect the predominant process or characterize desertification as a complex of estimates.

In Turkmenistan, the basic types of desertification include degradation of the vegetative cover, wind erosion, water erosion, water logging of rangelands, salinization of irrigated farmlands and formation of solonchaks due to the closure of Kara-Bogaz-Gol Bay of the Caspian Sea. We estimated desertification using a three classes scale.

Table 5.1. The criteria for assessment of vegetation degradation (Kharin, Nechaeva & Nikolaev et al., 1985).

Assessment factors	Class limits		
	Slight	Moderate	Severe and very severe
Plant communities	Climax or slightly changed	Long time existing	Ephemeral or shortly existing
Decrease in productivity, %			
Sandy desert	<15	15 – 35	>35
Gypsum desert	<20	20 – 40	>40
Clay desert	<35	35 – 70	>70
Foothill loess desert	<20	20 – 40	>40
Low mountain belt	<20	20 – 40	>40
Middle and high mountain belts	<15	15 – 30	>30
Decrease in area covered by climax vegetation			
Sandy desert	<10	10 – 35	>35
Gypsum desert	<10	10 – 50	>50
Clay desert	<20	20 – 45	>45
Foothill loess desert	<20	20 – 40	>40
Low mountain belt	<10	10 – 50	>50
Middle and high mountain belts	<5	5 – 40	>40

Table 5.2. The dynamics of plant associations by desertification in different types of desert (Kharin, Nechaeva &. Nikolaev et al., 1985).

Desertification classes	Plant associations	Number of registered plant species
	Sandy desert	
Zero level	*Haloxylon persicum – Carex physodes*	42
Slight	*Haloxylon persicum – Stipagrostis pennata + Carex physodes*	36
Moderate	*Salsola richtery – Stipagrostis pennata*	17
Severe	*Stipagrostis karelinii – Bromus tectorum*	8
Very severe	*Stipagrostis karelinii*	3
	Clay desert	
Zero level	*Salsola gemascens + Artemisia kemrudica – Gamanthus gamocarpus*	35
Low	*Salsola gemascens + Artemisia kemrudica – Gamanthus gamocarpus*	25
Moderate	*Artemisia kemrudica + Salsola gemascens*	11
Severe	*Artemisia kemrudica + Climacoptera lanata*	8
Very severe	*Peganum harmala*	3
	Foothill loess desert	
Zero level	*Carex pachystilis +Poa bulbosa*	50
Slight	*Carex pachystilis +Poa bulbosa*	47
Moderate	*Poa bulbosa + Carex pachystilis*	30
Severe	*Poa bulbosa + Iris songarica*	26
Very severe	*Peganum harmala*	3

Table 5.3. The criteria of assessment wind erosion in sandy desert (Kharin, 1994).

Assessment factors	Class limits		
	Slight	Moderate	Severe and very severe
Area covered with moving sand dunes, %	15 – 30	30 – 70	>70
Sod cover, %	30 – 50	10 – 30	<10
Vegetation density, %			
Shrubs	10 –15	5 – 10	<5
Herbaceous vegetation	40 – 65	10 – 40	<10

Table 5.4. Criteria for assessment of water erosion (Kharin, 1994).

Assessment factors	Class limits		
	Slight	Moderate	Severe and very severe
Types of water erosion	Sheet erosion (single cavities)	Sheet erosion (up to10 cavities per 1 km^2, single gullies)	Gully erosion (more than 10 cavities per 1km^2)
Ablation of surface soil layer, cm	<5	5 – 20	>20
Vegetation density, %			
Trees and shrubs	<20	20 – 50	>50
Herbaceous vegetation	<20	20 – 50	>50

Criteria for assessment of vegetation degradation

Degradation of the vegetative cover is the most common type of land degradation in Turkmenistan. The criteria of assessment are given in Table 5.1 (Kharin, Nechaeva & Nikolaev et al., 1985). Decrease of the area covered with climax vegetation was estimated by comparison of thematic maps from 1965 till 1990.

The dynamics of plant associations under the impact of desertification is shown by Table 5.2. The Table also indicates the decrease of species number by desertification.

A special case of plant cover degradation (undergrazing) was registered in the Karakum desert. This process is caused by formation of a thick layer of desert moss (*Tortula desertorum*). In some desert areas where there is no grazing due to the absent of water sources, this moss covers up to 40% of the surface, making an obstacle for regeneration of forage plants. That leads to reduction of biological productivity of desert ecosystems. Moderate grazing, on the other hand, stimulates plant reproduction because animals break the moss cover and loosen the soil.

Criteria for assessment of wind erosion

Wind erosion is the second important process of land degradation in Turkmenistan. Desertification as an anthropogenic process is combined here with natural physical movement of sands in the desert. Scientists of Turkmenistan Ivanov (1972) and Dobrin (1964) studied the movement of sand dunes in the desert. According to these studies sand particles smaller than 0.04 mm can be carried at great distances by the wind. Particles from 0.04 to 2.0 mm in size, however, are transported by the wind in a suspended conditions along the sand surface. These movable eolian landforms can be either formed in situ or transported from elsewhere. The surface layer of atmosphere contains layers with various whirlpool structures, and 97% of the sand is transported by the wind within 15 cm from the surface. In sandy deserts, deflation is not caused by human activities, such movable landforms occupy here 15% from the total area. The criteria

Table 5.5. Criteria for assessment of waterlogging rangelands (Kharin, 1994)

Assessment factors	Class limits		
	Slight	Moderate	Severe and very severe
Density of hydrophilous vegetation			
a. *Tamarix ramosissima, Alhagi persarum*	<30	-	-
b. *Tamarix ramosissima, Alhagi persarum and Karelinia caspia*	-	30 – 70	-
c. *Phragmites australis, Glycyrrhiza glabra and Alhagi persarum*	-	-	-
Depth of fresh or low mineralized ground water	5 – 10	2 – 5	<2
Soil humidity regime	Automorphic	Semi-hydromorphic	Hydromorphic

Table 5.6. Criteria for assessment of salinization of irrigated farmlands (Kharin, 1994).

Assessment factors	Class limits		
	Light	Moderate	Severe and very severe
Degree of salinization			
a. Total solid residue, %	0.210 – 0.400	0.400 – 0.600	>0.600
b. Cl^-, %	0.01– 0.030	0.030 – 0.100	>0.100
c. Na^+, %	0.023 – 0.046	0.046 – 0.092	>0.092
Mineralization of ground water, g/l	3 - 6	6 – 10	10 – 30
Mineralization of irrigation water, g/l	0.5 – 1.0	1.0 – 1.5	<1.5
Decrease in row cotton yield (% of the zero level)	<15	15 – 40	40 – 80
Seasonal salt accumulation			
a. Percentage	0.11 – 0.20	0.20 – 0.30	0.30 – 0.60
b. Metric tons per 1 ha	16 – 30	30 – 45	45 – 90
Degree of pollution of irrigation water (ratio of content of toxic chemicals to their allowed concentration)	1.0 – 6.0	6.0 – 11.0	>11.0

of assessment of wind erosion are given in Table 5.3 (Kharin,1994).

Criteria for assessment of water erosion

The criteria for assessment of water erosion are given in Table 5.4 (Kharin,1994).

Water erosion is mainly registered in mountains and foothills of Turkmenistan.

Criteria for assessment of waterlogging rangelands

The criteria for assessment of waterlogged rangelands are given in Table 5.5 (Kharin,1994). This process occurs due to discharge of irrigation water from oases to surrounded areas. Productivity of plant in this case is, as a rule, increased, but plant succession takes place, and forage plants are followed by non-palatable plants.

Criteria for assessment of soil salinization

Salinization of irrigated farmlands caused by overirrigation and improper agricultural technology cause the great economic damage in Turkmenistan (See Chapter 8 of this book). Cotton is a monoculture dominating on irrigated farmlands. As a result, the upper soil horizon becomes compact, the content of humus is reduced, aeration of soils is worsened, soil fauna degrades and number of microorganisms in soil is diminished. All that leads to reduction of cotton yield. 70% of soils in the irrigation zone have a low fertility because of low content of humus, nitrogen and phosphorus. The criteria for assessment of this type of desertification are given in Table 5.6.

Table 5.7. Inherent risk of desertification by assessment of land degradation in Turkmenistan.

Assessmnet factors	Class limits			
	Slight	Moderate	Severe	Very severe
Degradation of the vegetative cover				
Mechanical soil composition	Loam	Sandy loam	Loamy sand	Sand
Life form of dominant plants	Trees	Shrubs	Semi-shrubs	Grass
Wind erosion				
Mechanical soil composition	Loam	Sandy loam	Loamy sand	Sand
Mean annual wind speed, m/sec	2 – 5	5 – 10	10 – 15	>15
Percentage of active winds, per annum	< 5	5 – 10	10 – 15	>15
Water erosion				
Mechanical soil composition	Loam	Sandy loam	Loamy sand	Sand
Slopes, in grad.	< 5	5 – 10	10 – 15	>15
Salinization of irrigated farmlands.				
Salinity of ground water, g/l	3 – 6	6 – 10	10 – 20	>20
Salinity of irrigation water, g/l	0.5 – 1.0	1.0 – 1.5	>1.5	

Criteria for assessment of inherent risk of desertification

Above that, classification of inherent risk for Turkmenistan territory was developed (Table 5.7).

5.2. Remote sensing indicators of desertification

Spectral reflectance of desert landscapes and optimal seasons of remote sensing

Remote sensing technology is now widely used for desertification mapping. Optical properties of landscape elements include albedo, spectral reflectance coefficients (SRC) and indicatrix. Communities of desert plants have low density (crown cover) and the integral spectral reflectance is low (Radziminsky, 1975). By desertification, optical properties of plant cover are changed. A special experiment was conducted by the author in central Karakum in 1967. Five experimental sample plots, 400 m^2 in size, were taken and SRF of plants and sand surface were measured. The characteristics of these plots are given below Table 5.8). Plant cover was mapped in each plot and field measurements of SRC were conducted.

The results of spectrometric measurements are given in Table 5.9. Single measurements of SRC of different objects were conducted and the integral spectral reflectance for each five spectral bands was calculated. One can see that spectral band 600 – 700 mkm is the most informative for interpretation of desertification classes. Moving sand dunes and desertification spots around desert wells or settlements are always pictured by light photographic tone.

Water erosion in the arid zone mainly occurs in mountain and foothill country. During the process of water erosion, soil material is washed away and transported down the slopes. So the sites of soil washing are pictured by light tone, and the sites of accumulation of soil material by darker tone.

Some technogenic objects in the desert are characterized by optical properties which are close to that of natural objects. For example, drilling solution, mixed with oil products can produce artificial lakes around drilling stations by oil

Table 5.8. Percentage of plant cover in experimental plots of different desertification status.

No	Life forms of plants	Desertification classes				
		Zero level	Slight	Moderate	Severe	Very severe
1.	Shrubs Ephemera Moving sand	31 69 -				
2.	Shrubs Ephemera Moving sand		16 76 8			
3.	Shrubs Ephemera Moving sand			15 58 27		
4.	Shrubs Ephemera Moving sand				6 69 25	
5.	Shrubs Ephemera Moving sand					5 10 85

Table 5.9. Integral spectral reflectance of plant communities with different degree of desertification.

Number of plots	Desertification classes	Spectral bands, mkm			
		400 – 500	500 – 600	600 – 700	700 – 800
1	Zero level	0.08	0.177	0.145	0.127
2	Slight	0.07	0.106	0.134	0.151
3	Moderate	0.073	0.109	0.163	0.188
4	Severe	0.102	0.143	0.177	0.207
5	Very severe	0.114	0.192	0.254	0.275

Table 5.10. Phenologcal indicators of aerial and space photography in different physical regions of Turkmenistan.

Physical regions	Seasonal aspect	Phenological indicators	
		Beginning	End
Karakum desert	Spring	*Leafing out Haloxylon persicum*	Withering up *Carex physodes*
	Autumn	Formation of seeds of *Haloxylon persicum*	Full ripening of seeds of *Haloxylon persicum*
Badkhyz highland	Spring	Flowering of *Roemeria refracta*	Withering up *Carex pachystilis*
Oases	Summer	Barley seeds becoming juicy	Full ripening of wheat
	Autumn	Opening up of cotton bolls	Beginning of leaf coloration of apricot tree

Table 5.11. SRC of plants leaves of which were damaged with cement dust (Kiril'tseva, 1986).

Plant Names	Wave length, mkm								
	400	450	500	550	600	650	700	750	800
Ailant altissima									
1	0.05	0.07	0.08	0.12	0.05	0.05	0.21	0.52	0.51
2	0.03	0.05	0.13	0.14	0.13	0.15	0.30	0.45	0.46
Thuya oxiden talis									
1	0.05	0.06	0.05	0.11	0.04	0.03	0.28	0.39	0.40
2	0.04	0.04	0.07	0.09	0.06	0.05	0.20	0.32	0.34
Ulmus campestris									
1	0.04	0.04	0.09	0.14	0.08	0.04	0.08	0.45	0.52
2	0.05	0.05	0.0.09	0.11	0.11	0.09	0.15	0.36	0.44
Tamarx sp.									
1	0.05	0.04	0.10	0.15	0.10	0.06	0.34	0.48	0.52
2	0.03	0.04	0.08	0.16	0.16	0.14	0.29	0.38	0.41
Sophora japonica									
1	0.05	0.06	0.08	0.15	0.10	0.05	0.13	0.53	0.56
2	0.08	0.08	0.11	0.18	0.16	0.15	0.25	0.39	0.43
Maclura aurantica									
1	0.07	0.08	0.09	0.14	0.09	0.05	0.35	0.51	0.52
2	0.10	0.10	0.11	0.15	0.14	0.11	0.31	0.35	0.36
Acer negundo									
1	0.04	0.04	0.06	0.13	0.08	0.03	0.26	0.52	0.53
2	0.05	0.05	0.09	0.13	0.12	0.09	0.20	0.37	0.39

prospecting works. This technogenic surface has SRC close to that of takyr, but its indicatrix has a mirror effect, so these two objects can be distinguished only by large angles of registration by remote sensor.

As known, SRC of natural objects are changed during the seasons of the year. Phenological indicators of proper seasons of photography of other lands capes are given in Table 5.10.

During these seasons the contrast between various natural objects is higher to be compared with other seasons.

Spectral reflectance of desert plants is changed under the impact of atmospheric pollution or by damage of biological agents, like insects, fungi or microorganisms (Kiril'tseva, 1986). The SRC of some trees and shrubs of Turkmenistan damaged by cement dust in the vicinity of cement producing plants are given in Table 5.11. In the Table plants are placed in the sequence from slightly damaged to severely damaged. That can be explained by an individual reaction of plant species. *Ailant altissima* is the most resistant and *Acer negundo* is the most unstable. One can see that in the visible spectrum the SRC of damaged plants are 2 – 3 times higher than SRC of healthy plants. In the infrared spectrum, the SRC of damaged plants fall by 15 – 30%. Such changes can be registered by remote sensors.

Specifications of remote sensing for desertification monitoring

Specifications of remote sensing for monitoring desert environment are given in Table 5.12. Table 5.12 contains spectral bands, season of photography and ground resolution of imagery. Sun elevation above the earth is also indicated. Low sun angles improve the interpretative quality of photos by study of flat desert terrain.

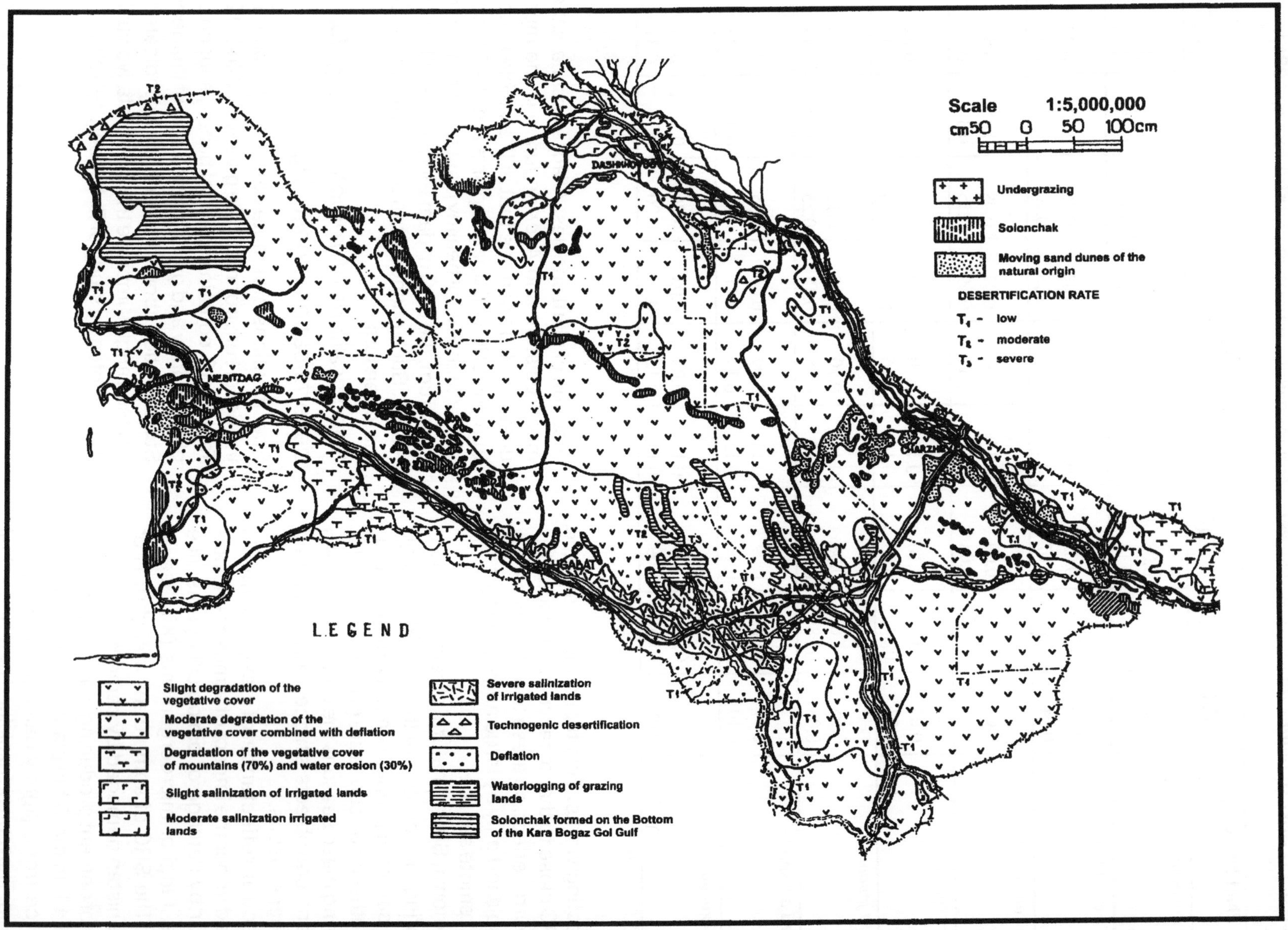

Figure 5.1. Desertification map of Turkmenistan.

Table 5.12. Spectral bands, seasons and ground resolution of space imagery by monitoring desert lands (Kharin, Kalenov & Kurbanmuradov et al., 1978, modified).

Tasks	Season, month	Sun elevation, degrees	Periodicity, months	Spectral bands, mkm	Ground resolution, m
Study of desert landscapes					
Regionalization and map compilation	April	20 – 30	120	600 – 700 750 – 850	50
Study of geomorphology	April	20 – 30	60	600 – 700 750 – 850	50
Study of human-induced changes	April	20 – 30	24	600 – 700 750 – 850	50
Study of soil cover					
Map compilation	April	20 – 30	120	600 – 650	10
Study of soil salinization	April	20 – 30	36	500 – 550 700 – 800	10
Study of the vegetative cover					
Map compilation, study of biological productivty	April, October	20 – 30	60	400 – 500 600 – 675	10
Study of phenology	January – December	20 – 30	1	600 – 700	50
Study of water resources					
Study of surface runoff, water reservoirs and canals	January – December	20 – 30	1	600 – 675 700 – 1100	10
Hydrological mapping	April	30 – 40	24	600 – 675 700 – 1100	10
Study of desert rangelands					
Rangeland mapping	April	20 – 40	60	400 – 525 600 – 675	10
Study of rangeland status	April	20 – 40	12	400 – 500 600 – 675	50
Study of soil erosion					
Wind erosion	June – September	20 – 30	12	500 – 650	10
Water erosion	October – March	20 – 40	12	700 – 1100	10

5.3. Compilation of desertification map of Turkmenistan

Desertification map of Turkmenistan

By desertification mapping, terrain mapping units should be identified and shown in the map. According to the concept, discussed in Section 5.1 of this book, Elementary Mapping Unit for desertification assessment (EMU) is a peace of terrain with similar characteristics of IRD and DS. Depending on mapping scale, EMU can correspond to different l andscape units.

Indian scientists (Iqbaluddin Saifuddun, Akram Javed & Monti, 1999) have developed the concept of Terrain Mapping Unit (TMU). It was an attempt to integrate the natural resource endowments of lithology, soil, ground water and geomorphology into TMU. As the authors of this approach emphasize, these natural entities make up the environmentally compatible units for land system management in agriculture and forestry. TMUs are management units, and the authors have proposed to use them instead of using artificial (administrative) and cultural boundaries. This concept is very close to the concept of Russian geographers on application of landscape units in thematic mapping.

A desertification map of Turkmenistan, compiled by the author is given in Figure 5.1. The small scale desertification map gives a general picture of the distribution of desertification processes in the country. Especially strong human influence is manifested in oases and in adjacent lands. Extremely high soil salinization is demonstrated, for example in the Tedzhen oasis. Another ecological danger zone is the dried-up Kara-Bogaz-Gol Bay of the Caspian Sea. The lowest level of desertification is recorded along the Turkmenistan border with Iran and Afghanistan due to restrained development of this area. Below is given a structure of the database of desertification developed by the author (Kharin, 1994).

Database on desertification

I. Causes of desertification (C).

CN – natural causes; CN a – air temperature, CNb – albedo, CNc – precipitation, CNd – air humidity, CNe – wind speed, CNf – dust storms, CH – anthropogenic causes; CHa – centralized figures of planned economy, CHb – control figures of economic development of administrative regions, CHc – reported data on fulfillment of economic plans by administrative regions, CHd – population density, CHe – area of deforestation, CHf – animal pressure on rangelands, CHg – area not used for crop cultivation, CHg1 – area occupied by roads, CHg2 – area used for residence and construction, Chg3 –area used for industry, CHh – ploughed area, CHI – irrigated area.

2. The process of desertification (D).

DV - degradation of the vegetative cover, DVa – plant species composition, DVb – productivity, DVc – forested area, DVd – forest stock, DVe – rangeland area, DVf – rangeland productivity, DVg – area of dried forests, DVh – disappeared plant species. DS – soil degradation; DSa – area under wind erosion, DSb – area under water erosion, DSc – area with decreasing content of humus,

DCd – area with hard soil pans, DSe – area with salinization and water logging; DE – deterioration of water resources; DEa – river debit, DEb – lake water volume, DEc –amount of surface runoff, DEd – level of ground water, DEe – quality of surface and ground water. DA – changes in animal population; DAa – animal species composition, DAb – disappered animal species, DAd – diseases of animals.

3. Consequences of desertification (A).

A S – social consequences; ASa –number of nomadic population, ASb – migration of population, ASc – new settlements, ASd – sanitary conditions of people, ASe – birth rate, ASf – death rate, ASg – specific human diseases, ASh – life span. AE – economic sequences of desertification; AEa – structure of sown crops, AEb – yield of agricultural crops, AEc – livestock population, AEd – food consumption, AEe – water consumption, AEf – fuel consumption, AEg – per capita income.

4. Measures to combat desertification (L).

LS – desertification monitoring; LSa – criteria of desertification, LSa1–criteria of current status, LSa2–criteria of current rate, LSa3 – criteria of inherent risk, LSa4 – grazing pressure, LSa5–density of population, LSa6 – methodology of desertification assessment. LM – measures to combat desertification; LMa – areas where measures on wind control were undertaken, LMb – areas where measures on soil salinization control were undertaken, LMc – areas where measures on control of soil compaction and crusting were undertaken, LMd – areas of pasture melioration were undertaken, LMe – areas of measures on melioration of arable lands were undertaken,

LMf – meliorated forest areas, LMg – usage of mineralized water, LMh1 – deposit of mineralized water, LMh2 – area of irrigated crops, LMh3 – yield of irrigated crops, LMi – areas where complex measures on desertification control were undertaken. LP – forecast of desertification, LPa – desertification models, LPb – climate forecast, LPc–population density forecast, LPd – economic development forecast.

5. Archive materials (A R).

ARa – topographic maps, ARb – thematic maps, ARc – aerial photos, ARd – space photos, ARe – field descriptions, ARf – bibliography.

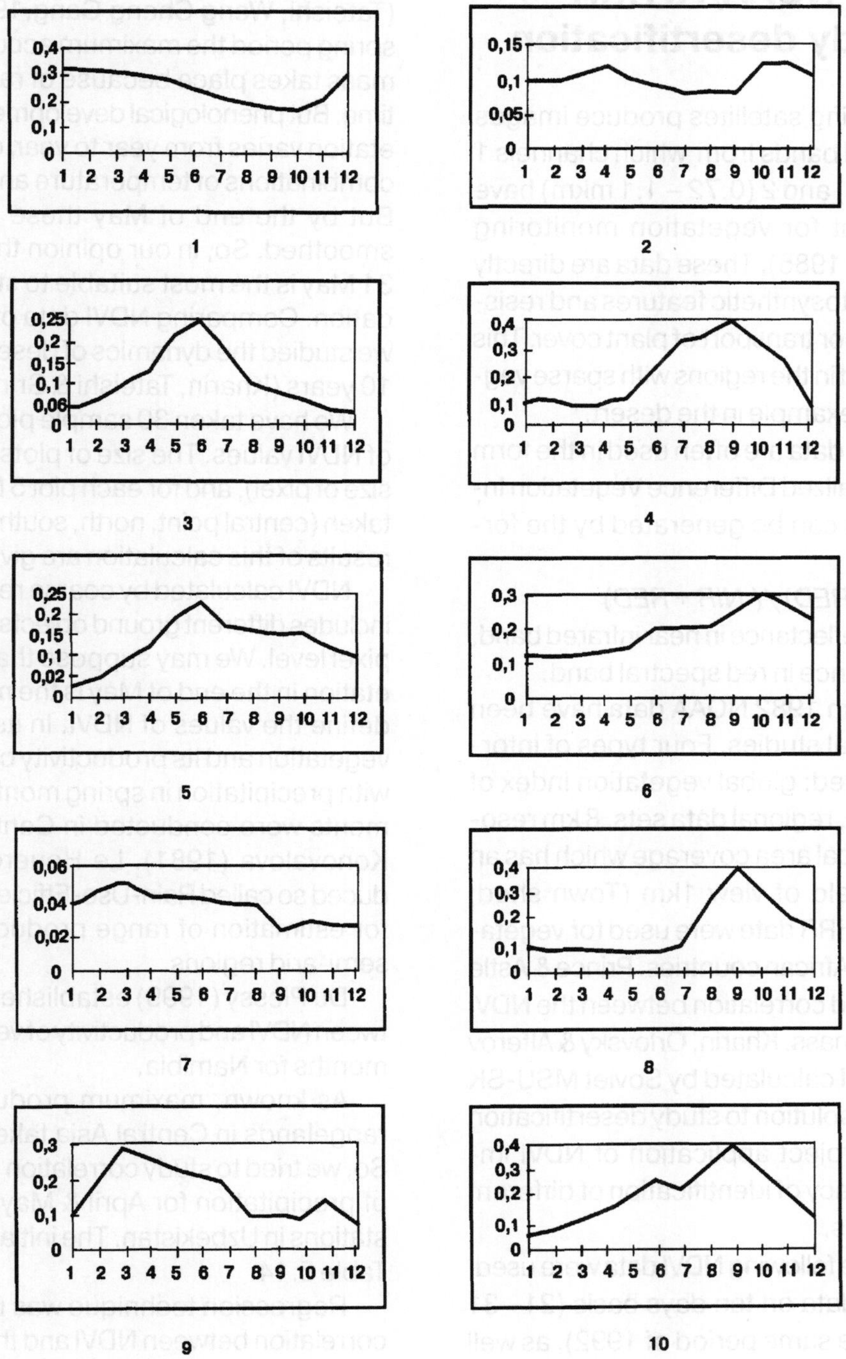

Figure 5.2. Monthly dynamics of NDVI for main types of landscape in Central Asia (Tateishi & Wen Cheng Gang, 1995). 1 – cespitose grass steppe, 2 – northen desert with Artemisia sp., 3 – northen semi-desert with semi-shrub vegetation, 4 – southern sandy desert with Haloxylon sp., 5 – southern desert with Artemia sp., 6 – northen desert with shrub vegetation, 7– southern desert with psammophylous vegetation, 8 – semisavanna on foot-hill plain, 9 – moving sand dunes, 10 – oases with cultural vegetation.

5.4. Use of NOAA/AVHRR data to study desertification

NOAA polar-orbiting satellites produce images in several spectral bands from which channels 1 (0.55 – 0.68 mkm) and 2 (0.72 – 1.1 mkm) have a special interest for vegetation monitoring (Tucker & Sellers, 1985). These data are directly related to the photosynthetic features and resistance of water vapor transport of plant cover. This effect is weakened in the regions with sparse vegetation cover, for example in the desert.

NOAA/AVHRR data are often used in the form of so called Normalized Difference Vegetation Index (NDVI) which can be generated by the formula:

$NDVI = (NIR - RED)/(NIR + RED)$

where: *NIR* – reflectance in near infrared band,

RED – reflectance in red spectral band.

Beginning from 1982 NOAA data have been using in ecological studies. Four types of information can be used: global vegetation index of coarse resolution, regional data sets, 8 km resolution data and local area coverage which has an instantaneous field of view 1km (Town shed, 1994). NOAA/AVHRR date were used for vegetation monitoring in African countries. Prince & Astle (1986) established correlation between the NDVI values and phytomass. Kharin, Orlovsky & Alferov (1990) used NDVI calculated by Soviet MSU-SK photos, 160 m resolution to study desertification in Mali. In this project application of NDVI improved the accuracy of identification of different land cover classes.

In our study the following NDVI data were used: 8 km resolution data on ten days basis (21– 31 May 1982 and the same period of 1992), as well as 1 km resolution data for 15 meteo-stations of Uzbekistan, 1982 and 1992. The region of Central Asia within the following limits was studied: 55 – 67° E and 50° N – southern borders of Central Asian countries.

As known spring is the most contrast season in Central Asia. Monthly dynamics of NDVI (8 km) resolution measured in 1990 is given in Figure 5.2 (Tateishi, Weng Cheng Gang, 1995). During the spring period the maximum accumulation of biomass takes place because of rainfall during this time. But phenological development of desert vegetation varies from year to year, depending upon combinations of temperature and precipitations. But by the end of May these differences are smoothed. So, in our opinion the period of 21 – 31 May is the most suitable to study land degradation. Comparing NDVI data of 1982 and 1992 we studied the dynamics of desertification during 10 years (Kharin, Tateishi & Gringof, 1998).

We have taken 30 sample plots for calibration of NDVI values. The size of plots was 1x1 km (the size of pixel), and for each plot 5 NDVI values were taken (central point, north, south, west, east). The results of this calculation are given in Table 5.13.

NDVI calculated by coarse resolution imagery includes different ground objects clustered at subpixel level. We may suppose that density of vegetation in the end of May is the main factor which define the values of NDVI. In its turn, density of vegetation and its productivity closely correlated with precipitation in spring months. Such experiments were conducted in Central Asia by An & Konovalova (1981). Le Houerou (1994) introduced so called Rain-Use-Efficiency Factor (RUE) for estimation of range productivity in arid and semi-arid regions.

Du Plessy (1999) established correlation between NDVI and productivity of vegetation in spring months for Namibia.

As known, maximum productivity of desert rangelands in Central Asia take place in spring. So, we tried to study correlation between the sum of precipitation for April & May and NDVI for 15 stations in Uzbekistan. The initial data are given in Table 5.14.

Regression technique was used to establish correlation between NDVI and the sum of precipitation. The equations are given below:

$y = 88.10 - 177.3x,\ r = 0.62$ (1982)

$y = 381.01 - 622.65x + 622.67x2,$

$r = 0.726$ (1992)

where:

y – the sum of precipitation, April & May

x – NDVI, May 21 – 31

r – correlation coefficient.

Table 5.13. NDVI data for different land cover classes of Central Asia (Kharin, Tateishi & Gringof, 1998).

No	Longitude, deg., min.	Latitude, deg., min.	NDVI 21 – 31 May, 1982	NDVI 21 – 31 May 1992	Land cover classes
1	57.30	41.00	0.2 – 0.25	0.25 – 0.3	Open shrubland of sandy desert
2	59.25	41.00	0.3 – 0.35	0.25 – 0.3	Agricultural land (cotton, gardens)
3	62.10	42.50	0.2 – 0.25	0.25 – 0.3	Open shrubland of sandy desert
4	65.10	43.10	0.2 – 0.25	0.3 – 0.35	Open shrubland of sandy desert
5	62.10	41.50	0.2 – 0.25	0.3 – 0.35	Open shrubland of sandy desert
6	66.30	40.10	< 0.35	> 0.35	Agricultural land (cotton)
7	65.30	38.50	< 0.35	> 0.35	Agricultural land (cotton)
8	67.40	39.45	< 0.35	> 0.35	Uniperus forest
9	63.30	36.20	0.3 – 0.25	> 0.35	Semisavanna
10	56.20	40.30	0.2 – 0.25	0.25 – 0.3	Open shrubland
11	55.00	41.30	0.2 – 0.25	0.2 – 0.25	Open semishrubland
12	55.00	40.30	0.2 – 0.25	0.15 – 0.2	Open semishrubland
13	56.35	41.00	0.15 - 0.2	0.15 – 0.2	Solonchak
14	54.10	38.30	0.1 – 0.15	0.15 – 02	Open shrubland
15	54.30	38.00	0.1 – 0.15	0.1 – 0.15	Takyr
16	57.25	38.10	0.2 – 0.25	>0.35	Juniperus forest
17	54.45	39.00	0.1 – 0.15	0.15 – 0.2	Takyr
18	60.20	37.30	< 0.35	0.3 – 0.35	Agricultural land (cotton)
19	59.30	39.20	0.15 – 0.2	0.2 – 0.25	Shrubland in sandy desert
20	58.10	40.05	0.2 – 0.25	0.25 – 0.3	Shrubland
21	57.40	39.20	0.15 – 0.2	0.15 – 0.2	Degraded shrubland
22	56.30	39.30	0.1 – 0.15	0.15 – 0.2	Degraded shrubland
23	59.28	38.25	0.15 – 0.2	0.2 – 0.25	Secondary shrub community
24	63.30	38.20	0.2 – 0.25	0.2 – 0.25	Desert shrubland
25	62.15	37.10	0.25 – 0.3	0.25 – 0.3	Secondary shrub community
26	63.40	36.30	0.3 – 0.35	> 0.35	Semisavanna
27	61.30	35.50	< 0.35	> 0.35	Pistacia vera forest
28	61.10	36.30	< 0.35	0.3 – 0.35	Jungle in river valley
29	62.25	38.50	0.2 – 0.25	0.2 – 0.25	Moving sand dunes
30	58.30	37.50	0.25 – 0.3	> 0.35	Deciduous mountain forest

Finely, we tried to use NDVI values (with the interval of 0.05) to study desertification in Central Asia. For this purpose we calibrated NDVI values on existing thematic and desertification maps of Central Asia. The results of desertification monitoring are given in Table 5.15.

The territory of northern Kazakhstan was excluded from assessment because of low inclination of the satellite's orbit. The change of the area of moving sand dunes (Class 1) shows the dynamics of desertification during 10 years. The areas of land cover class 3 reduced during the same period probably because better moisture conditions in 1992. Expansion the size of oases (Class 7) was the result of human activities.

Table 5.14. Two months precipitation and NDVI values on May 21 – 31 in Uzbekistan (Kharin, Tateishi & Gringof, 1998).

No	Stations	Sum of precipitation, April & May, mm	NDVI, May 21 – 31
1	Chabankazgan		
	1982	No observ.	0.175
	1992	120	0.225
2	Akbaital		
	1982	134	0.215
	1992	167	0.325
3	Kulkuduk		
	1982	134	0.195
	1992	No observ.	0.325
4	Tamdy-Bulak		
	1982	125	0.225
	1992	135	0.315
5	Buzaubai		
	1982	135	0.225
	1992	185	0.325
6	Ayakagitma		
	1982	No observ.	0.235
	1992	132	0.295
7	Mashikuduk		
	1982	137	0.285
	1992	135	0.325
8	Karakalpakia		
	1982	130	0.225
	1992	133	0.245
9	Zhaslyk		
	1982	114	0.225
	1992	116	0.275
10	Nurata		
	1982	146	0.35
	1992	219	0.36

The experiment testifies to the fact that coarse resolution space imagery can be used for desertification monitoring at a regional level. Better results can be obtained if 1 km resolution is used (Kharin, Tateishi & Harakhsheh, 1999). For calibration of NDVI more perfect system of ground truth collection should be developed.

5.5. Multistage desertification monitoring

According to CCD desertification monitoring system should be created to study all changes occurred in desert environment. In our opinion, this system will include six levels of assessment: global, regional, sub-regional, national, district and local (Table 5.16).Global assessment includes the annual summarizing of regional data collected by agencies functioning at lower levels. According to CCD, Asia is considered as one region, so desertification map compiled by Kharin, Tateishi & Harahsheh 1999) corresponds to the regional level of desertification assessment. The countries of Central Asia, from this point of view belong to one sub-region.

The monitoring at the levels A, B, C and D is based on application of space imagery, at the level F aerial photos should be used. For realization of this system, a special infrastructure should be developed.

Several technical systems of remote sensing are used at present time for investigation of natural resources. First of all the space systems should be mentioned. The main of them are as following (Gabruk & Gershenzon, 1997; Dmitriev, Murakhtanov & Sukhikh, 1989):

1. American space system of investigation of natural resource Landsat. The first Earth Resources Technology Satellite was launched to the orbit in 1972. The next series of this satellite were called Landsat. Landsat 5 functioning at the orbit at present time has a special device Thematic Mapper (TM) which can register the image in seven spectral bands (mkm): 0.45 – 0.52, 0.52 – 0.60, 0.63 – 0.69, 0.76 – 0.90, 1.55 – 1.75, 2.08 – 2.35 and 10.40 – 12.00. TM has ground resolution 30 m in visible and near infrared bands and 120 m in thermal infrared band.

2. French space system SPOT(System Provisoire d'Observation de la Terre) has been functioning since 1986. SPOT 4 has special device HRVIR (High Resolution Visible and Infrared). By multispectral sensing the image can be obtained in the following spectral band (mkm): 0.50 – 0.59, 0.61 – 0.68, 0.79 – 0.89, 1.58 – 1.75, the ground resolution being 20 m. By photographing in one

Table 5.15. Dynamics of the areas affected by desertification in Central Asia during 1982 – 1992, in percentage (Kharin, Tateishi & Gringof, 1998).

No	Land degradation classes	NDVI	Percentage of area		Changes, %
			1982	1992	
1	Moving sand dunes	<0.1	9.2	9.7	-0.5
2	Solonchak and shallow lakes	0 – 0.05	20.7	18.2	- 2.5
3	Severely degraded desert land with herbaceous vegetation	0.05 – 0.1	17.4	15.2	- 2.2
4	Severely degraded desert land with shrub and semishrub vegetation	0.10 – 0.15	8.6	10.2	+0.6
5	Moderately degraded desert land with shrub and semishrub vegetation	0.15 – 0.20	6.9	6.4	- 0.5
6	Slightly degraded semisavanna and desert land with shrub and semishrub vegetstion	0.20 – 0.25	5.6	5.9	+0.3
7	Oases with cultural vegetation	0.25 – 0.30	4.7	5.4	+0.7
8	Mountain areas with joining oases	0.30 – 0.35	3.2	6.2	+3.0
9	Territory of the northern Kazakhstan excluded from assessment	>0.35	14.6	14.7	+0.1

Table 5.16. Multistage desertification monitoring system based on remote sensing.

Stages	Monitoring level	Area to be studied	Periodicity of observations	Ground resolution of the imagery	Scale of the final map
A	Global	Total	1 year	Regional desertification maps are summarized	1: 10,000,000
B	Regional	Total	1 year	8 km, 1 km	1: 5,000,000
C	Sub-regional	Total	1 year	1 km	1: 4,000,000
D	National	Total	1 year	1 km	1: 4,000,000 – 1: 2,500,000
E	Provincial	Selective	2-3 years or on a special occasion	100 m	1: 1,000,000
F	Local	Selective	3-5 years or on a special occasion	1 m	1: 100,000 – 1: 50,000

spectral band (0.61 – 0.68 mkm) ground resolution is 10 m.

3. Russian space system of investigation of natural resources RESURS is widely used in Russia. Earth satellites RESURS series have been using since 1974. For example Sukhikh, Kharin & Butusov (1999) used RESURS – 1 photos for mapping taiga forests. Scanner MSU – U installed in this satellite can produce photos in three spectral bands (mkm): 0.5 – 0.6, 0.6 – 0.7 and 0.8 – 0.9. The size of pixel is 35 x 45 m.

4. Meteorological space systems. In the former USSR the first satellite of Meteor was launched in 1969. Multispectral radiometer can obtain imagery in four spectral bands (mkm): 0.5 – 0.7, 0.8 – 1.0, 10.4 – 11.3 and 11.5 –

12.5 with ground resolution 1 km. Satellites of this series are used mainly to study cloud cover and other meteorological events.

5. NOAA space system. The first meteorological satellites on the program of NOAA, known as Tiros (Television and Infrared Observation Satellites) were launched in the USA in sixties. Kharin, Tateishi & Harahsheh (1999) used these pictures for desertification mapping in Asia. Application of this type of imagery to study desertification is discussed in Section 4.4 of this Book.

Aerial photos are also widely used for field mapping desertification and preparation of various thematic maps. Black-and-white, color and false color films are used to study natural resources on the desert. Panchromatic films, sensitive in 400 – 700 mkm spectral band are the most widely used films. Infrared films, sensitive to near infrared radiation (> 750 mkm), are also used, especially by interpretation of physical objects in the desert where mesophylous vegetation is common.

Color photographic films, sensitive to three spectral bands (blue, green and red) are also used. But its application is limited mainly to study special objects or special events, for example phenology of forests.

Of special interest are so called false color films, producing the imagery in "false" (not natural) colors.The main false color films used in Russia are indicated in Table 5.17.

Films SN-6 and SN- 23 give good results by studying of taiga forests and for interpretation of some physical objects in the arid zone (for example, river valleys, irrigated lands and oasis bordering areas). Films SN -4, SN-5 and in some cases SN–23 are suitable for interpretation of geological objects in the desert where the vegetative cover is not dense. They also can be used for desertification mapping. Film SN-3 doesn't have any practical application because the air haze smooth image by aerial photography.

Table 5.17. Characteristics of the main false color films (Savinykh, Kuchko & Stetsenko, modified, 1997).

Names	Filters	Wave lengths of maximum sensitivity, mkm	
		Upper layer	Low layer
SN-6m	Yellow	0.76	0.65
	Orange	0.76	0.64
	Red	0.74	0.64
SN-3	Without filter	0.45	0.64
SN-4	Yellow	0.55	0.66
SN-5	Yellow	0.74	0.55
SN-23	Yellow	0.74	0.55

5.6 Conclusion

Assessment of desertification is based on criteria developed for each type of land degradation (degradation of vegetation, wind erosion, water erosion, soil salinization etc.). As criteria can be used: the productivity, composition and density of vegetation, physical and chemical properties of soil, erosion features and other environmental changes caused be desertification.

Degradation of vegetation and soil cause as a rule, the change of spectral reflectance of land cover type. These changes can be registered by remote sensor in different spectral bands. Proper selection of spectral bands and the seasons of photography can improve the accuracy of desertification mapping.

NDVI can be used for assessment and mapping desertification because this criteria was found to closely correlated with land degradation. NOAA/AVHRR data are cheap sources for assessment of desertification in vast regions.

Desertification monitoring can be conducted at five levels: local, provincial, national, subregional,regional and global. The scales and ground resolution of photos are different for different levels of assessment.

CHAPTER 6

STATUS OF DEGRADED ARID LANDS

We must not wait for the mercy of nature,
To take it is our task.

I. Michurin

6.1. Change of biodiversity under the impact of desertification

Change of plant species composition

Degradation of drylands forms a threat to biological diversity and the carrying capacity of these lands. What is biological diversity? The Convention on Biological Diversity describes it as "The variability among living organisms from all sources including terrestrial, marine and other aquatic ecosystems and the ecological complexes which they are part, this includes diversity within species, between species and of ecosystems". Biodiversity is difficult to measure, especially to express quantitatively the changes that occurred under the impact of desertification. The study and mapping of biodiversity in drylands was neglected field until recently. Under the impact of desertification, biodiversity within ecosystems is reduced. This trend can be illustrated by Table 6.1 which gives the results of obser vations conducted by Nechaeva in central Karakum and published by Kharin, Nechaeva & Nikolaev et al. (1985).

Zero level characterizes the status of ecosystems which are not affected by desertification. On the final stage of plant cover degradation monodominant communities are formed, for example that of *Peganum harmala*, a poisonous plant. This process is leading to reduction of biomass and decrease of nutritive value of desert rangelands. The results of observations conducted by the same scientist in the central Karakum are given in Table 6.2.

In the mountains of Central Asia many plant species have been used by the people because of their valuable food or medical properties. Native people consider these plants as inexhaustible natural resources. For example, nuts of *Amygdalis communis* are consumed uncooked or are added to dough. Wild onions (*Allium regeli, A. fibrosum*) are consumed uncooked or are seasoned with soup. Rosette leaves of *Eremorus sp*. are used in salad. Leaves of *Rheum turkestanicum* have a nice sour taste and they have been consumed by people since ancient times. Spinach (*Spinacia turkestanica*) is used for stuffing patties. Among the wild food plants are: *Morus alba, Barberis integerrima, Eleagnus orientalis, Nitraria shoebery*, etc.

Table 6.1. Dynamics of associations and species number in different types of deserts (Kharin, Nechaeva & Nikolaev et al., 1985).

Classes of desertification	Associations	Species numbers
	Sandy desert	
Zero level	*Haloxylon persicum – Carex physodes*	42
Slight	*Haloxylon persicum + Stipagrostis pennata + Carex physodes*	36
Moderate	*Salsola richteri – Stipagrostis pennata*	17
Severe	*Stipagrostis karelinii + Bromus tectorum*	8
Very severe	*Stipagrostis karelinii*	3
	Gypsum desert	
Zero level	*Artemisia badhysi + Salsola orientalis – Carex pachystylis*	31
Slight	*Artemisia badhysi + Salsola orientalis – Carex pachystylis*	29
Moderate	*Ephedra distachia + Artemisia badhysi*	19
Severe	*Ephedra distachia – Horaninovia anomala*	14
Very severe	*Peganum harmala*	8
	Clay desert	
Zero level	*Salsola gemascens + Artemisia kemrudica – Gamanthus gamocarpus*	5
Slight	*lsola gemascens + Artemisia kemrudica – Gamanthus gamocarpus*	5
Moderate	*Artemisia kemrudica – Salsola gemascens*	11
Severe	*Artemisia kemrudika – Climacoptera lanata*	8
Very severe	*Peganum harmala*	3
	Piedmont loess desert	
Zero level	*Carex pachystylis + Poa bulbosa*	50
Slight	*Carex pachystylis + Poa bulbosa*	47
Moderate	*Poa bulbosa + Carex pachystylis*	30
Severe	*Poa bulbosa + Iris songarica*	26
Very severe	*Peganum harmala*	7

Table 6.2. Change of forage nutritive value of desert rangelands in central Karakum, in forage units per hectare (Kharin, Nechaeva & Nikolaev et al., 1985).

Classes of desertification	Spring	Summer	Autumn	Winter
	Association *Haloxylon persicum – Carex physodes*			
Zero leve	1 68	134	119	80
Slight	92	32	91	63
Moderat	103	76	17	7
Severe	73	36	16	6
	Association *Calligonum rubens, Mausolea eriocarpa – Carex physodes*			
Zero level	95	112	61 30	
Slight	219	13	6 60 35	
Moderate	100 83 45 20			
Severe	65 60 34 12			
	Association *Salsola arbuscula – Artemisia kemrudica – Carex physodes*			
Zero level	170	160	71	30
Slight	133	117	70	26
Moderate	109	50	38	18
Severe	70	45	28	14

Table 6.3. Rare and endangered plants of Central Asia (Red Data Book of the USSR 1978).

Name of species	1	2	3	4	5	Limiting factors
Species populations of which are probably disappeared						
Amygdalis vavilovii	-	-	-	+ -	-	Unidentified
Astragalus arianus	-	-	-	+	-	Narrow range, human impact
Dionysia kossinskyi	-	-	-	+	-	Specific habitat
Dorema karataviense	+ -	-	-	-	-	Specific ecological features of plant
Eremurus korovinii	-	-	+	-	-	Changes of habitat due to grazing
Gypsophila aulietensis	+	-	-	-	-	Stenotopic plant with narrow range, land development
Populus cataracti	-	-	-	+	-	Unidentified
Pterocephalus fruticulosus	-	-	-	+	-	Unknown
Rhamnus scravschanicus	-	-	+	-	-	Destruction of habitat by road construction
Salvia baldshuanica	-	-	+	-	-	Unknown
Salvia gontscharovii	-	-	+	-	-	Unknown
Sorbaria olgae	-	-	-	-	+	Unidentified
Viola hissarica	-	-	+	-	-	Very narrow range, agricultural development
Endangered species						
Allochrusa tadshikistanica	-	-	+	-	-	Cutting wood
Atropa komarovii	-	-	+	-	-	Change of habitat, grazing
Brionia lappifolia	-	-	+	-	-	Unknown
Eremurus kopetdaggensis	-	-	-	+	-	Water erosion, grazing
Eulophia turkestanica	-	-	-	+	+	Land reclamation
Jasminum revolutum	-	-	+	-	-	Low number, narrow range
Juno magnifica	-	-	-	-	+	Collecting plants, grazing
Lonicera paradoxa	-	+	+	-	-	Relict, cutting for fuel
Malus niedzwetzky	+	-	-	-	-	Intensive human activity
Mandragora turcomanica	-	-	-	+	-	Poor seed regeneration, damage of seeds by birds, destruction of fruits by man
Prunus darvasica	-	-	-	+	-	Unidentified
Sclerotia pentaceros	+	-	-	+	-	Narrow ecological range, grazing
Ungernia spiralis	-	-	-	+	-	Narrow ecological range, grazing

(Note : 1 – Kazakhstan, 2 – Kirghizstan, 3 – Tajikistan, 4 – Turkmenistan, 5 – Uzbekistan).

Plant species from The Red Data Book

The Red Data Book of the USSR (1978) includes 92 species of flora of Central Asia (Table 6.3). They are divided into four categories:

1. Species population of which has been probably disappeared (13 plant species),
2. Endangered species (13 plant species),
3. Rare species (46 plant species),
4. Species decreasing in number (14 plant species).

Conservation of biodiversity needs special programs to provide rehabilitation of rare species in situ or *ex-situ*. For example, only 300 individuals of *Mandragora turcomanica* are found in natural habitats of the Kopetdag Mountains (Kharin, 1995). National action program to combat desertification in Turkmenistan (1996) envisaged special measures on this subject. Study of biological features and the ecology of rare plants is the first step in this program. Difficulties arise by growing rare plants in botanical

Table 6.3 (continued)

Name of species	1	2	3	4	5	Limiting factors
			Rare species			
Abies semenovii	+	-	-	-	-	Cutting wood, grazing
Allium sergii	+	-	-	-	-	Narrow ecological change of habitats
Amigdalus susakensis	-	+	-	-	-	Destruction of habitat
Astragalus bobrovii	-	-	-	+	+	Grazing, cutting, tourism
Astragalus rubellus	-	-	-	-	+	Limited habitat
Astragalus rubtzovii	+	-	-	-	-	Relict species, grazing
Atraphaxis badghysi	-	-	-	+	-	Unknown
Atraphaxis muschketovii	-	-	-	+	-	Land reclamation
Atraphaxix terentifoliai	+	-	-	-	-	Land reclamation
Berberis iliensis	+	-	-	-	-	Land reclamation, regulation of river flow
Berberis karkarlensis	-	-	-	-	+	Berry collecting
Bryonia melanocarpa	-	-	-	-	+	Grazing
Calligonum triste	+	-	-	-	-	Cutting for fuel, grazing
Cephalantera longifolia	-	+	+	-	-	Plant collecting, destruction of habitat-
Cerasus blinovskii	-	-	-	+	-	Cutting, grazing
Colutea atabaevii	-	-	-	+	-	Grazing, cutting, unidentified factors, preventing germination of seeds
Cousinia grandifolia	+	-	-	-	-	Unidentified
Crucianella sintenisil	-	-	-	+	-	Unidentified
Delphinium oveczinnicovi	-	-	-	+	+	Low competitive capacity, cross breeding
Draba odudiana	-	-	+	-	-	Stenotopic plant, geological prospecting work
Ferula gigantea	-	-	+	-	-	Collecting latex, agricultural reclamation
Ferula turcomanica	-	-	-	+	-	Grazing
Hyacintella transcaspica	-	-	-	+	-	Bulb digging
Iris darvasica	-	-	+	-	-	Collecting flowers, economic activity
Iris ewbanhiana	-	-	-	+	-	Collecting flowers, digging out bulbs
Iskandera hissarica	-	+	-	-	-	Not studied
Kaufmania semenovii	+	-	-	-	-	Change of habitat
Komarovia anisosperma	-	-	-	-	+	Geological prospecting works, road construction
Kosopolianskia turkestanica	+ -	-	-	-	-	Narrow range, grazing

(Note : 1 - Kazakhstan, 2 - Kirghizstan, 3- Tajikistan, 4 - Turkmenistan, 5 - Uzbekistan).

gardens. In this case conservation of rare species should be conducted *in situ* (See Section 9.4 of this Book).

6.2. Degradation of forests

Classification and exploitation of forests in Central Asia

According to the classification of forests in the former USSR, the main part of Central Asian forests belonged to the first category – protective forests. General characteristics of forests of Central Asia are given in Table 6.4.

Official statistics on forestry in Central Asian Republics of the former USSR is giving in Tables 6.5 and 6.6.

Tables above testify to poor forest resources of Central Asia.The size of the forest area diminished during six years in all republics, except Kazakhstan. Table 6.6 needs some explanation. According to official statistics only the size of wood used for industrial purposes was indicated, but in Central Asia fuel wood was also cut. So, for example, in Turkmenistan reduction of the forest area was due to wood cutting for fuel.

The volume of wood cutting in the Republics of Central Asia is given in Table 6.7.

Table 6.3 (continued)

Name of species	1	2	3	4	5	Limiting factors
Lepidolopha fed-henkoana	+	-	-	-	-	Intensive grazing
Lonicera karataviensis	+	-	-	-	-	Cutting wood, economic development
Mogoltavia sewezowii	- +	-	-	-	-	Specific habitat
Ophris transhyrcana	-	-	-	- +	-	Destruction of habitat
Orchis fedchenkovii	-	-	-	+	-	Plant collection
Ostrowskia magnifica	-	-	-	+	+	Destruction of habitat, grazing, digging out roots
Populus berka rensis	+	-	-	-	-	Climatic changes, grazing, cutting
Pirus asia-media	-	-	-	-	+	Unidentified
Pirus cajon	-	-	+	-	-	Cutting, grazing
Rhaponticum aulieatens	+	+	-	-	-	Stenotopic habitat, human impact
Ribes malvifolium	-	-	+	-	+	Unknown
Sphaerosciadium denaese	-	-	+	-	+	Range is limited due to tourist activity and grazing
Zeravschaniaca zegeliana	-	-	+	-	-	Unknown
Zygophyllum bucharucim	-	-	+	+	+	Grazing, poor regeneration due to soil conditions
Species decreasing in number						
Allium vavilovii	-	-	-	+	-	Plant collection
Cladium mariscus	+	+	+	+	+	Land reclamation, pollution of water
Diospyrus lotus	-	+	+	-	-	Cutting, change of habitat
Eremurus aitchisonii	-	-	+	-	+	Mass plant collection
Euonimus koopmannii	+	-	-	-	-	Change of habitat due agricultural reclamation
Ferula sumbul	-	+	+	-	-	Low competitive capacity of plant
Ficus carica	-	-	+	+	+	Cutting, grazing, agricultural reclamation
Fritillaria eduardii	-	-	+	-	-	Plant and bulb collection
Fritillaria reddeana	-	-	-	+	-	Bulb and plant collection
Platanus orientalis	-	-	+	+	+	Cutting, grazing, soil compaction
Punica granatum	-	+	+	+	+	Intensive land development
Scorzonera tautagyz	+	+	+	+	+	Stenotopic habitat, low competitive capacity, collection of plants
Ungernia victoris	-	-	+	-	-	Digging out bulbs, collection of leaves
Vitis hissarica	+	-	+	-	-	Cutting, grazing, fires

(Note : 1 - Kazakhstan, 2 - Kirghizstan, 3- Tajikistan, 4 - Turkmenistan, 5 - Uzbekistan).

Volume of nuts and other products stored in forests

Wood is the main, but not sole product extracted from forests. Forests of Central Asia are rich in secondary product: nuts, food plants and medical plants. Data on storing up wild nuts in the forests of Central Asia, except of Kazakhstan, are given in Table 6.8. They included the nuts of *Juglans regia, Pistacia vera* and *Amigdalus* sp.

Wild plants are also used for collection of fruits and berries in the forests of Central Asia. The following plants are used as food plants: apple, pear, apricot, hawthorn, barberries, sweetbrier, raspberries and blackberries. The volume of wild fruits stored up in four countries of Central Asia is given in Table 6.9.

Several medical, food and technical plants are also collected in the forests of Central Asia.

Table 6.4. General characteristics of forests in Central Asia (Babaev & Kharin, 1994 a, Isaev 1991)

Countries	Land types	Forest types
Kazakhstan	High mountain forests (Northern Tian Shan, Dzhungarian Alatau, Ore Altai, Tartagai, Sary Sakmy, Central Altai)	Coniferous forests with *Larix sibirica, Picea abovata, Abies sibirica, Pinus sibirica*
	Low mountains (Ranges of Chingiz Tau, Kul'binsky, Manrak and foothills)	Broad-leaved forests (with dominance of *Malus silvestris, Crataegus* and other sp.)
	Plains with small tracts of mountains	Forests of *Betula* sp., Deciduous forests of Western Kazakhstan with dominance of *Quercus* sp., *Betula* sp., *Acer juglans, Betula* sp., *Salix* sp., etc.; Forests of river valleys (*Ulmus pumula*); Desert woodland: *Haloxylon* sp., *Salsola* sp., *Calligonum* sp.
Kirghizstan	High mountains (Kirghizssky Range, Ranges of Kungen Alatau, Tersky Alatau	Coniferou forests: *Picea schrenkiana, Abies sibirica*
	Middle elevation mountain (Fergansky Range)	Broad-leaved forests: *Acer turkestanicum, Acer semenovii, Malus kirghisorum, Malus silvestri, Juglans regia, Juniperus tuekestanica* and other sp.
	Accumulative plains	Broad-leaved forests with dominance of *Populus deversifolia, Betula* sp.
Tajikistan	Low mountains of Southern Tajikistan	*Xerophylous* woodland: *Pistacia vera, Amygdalus bucharica, Acer pubescens*
	Middle elevation mountains of Central Tajikistan (Southern slopes of Gissar Range, Ranges of Karateginsky, Vakhshky, Darvazsky)	Broad-leaved forests with *Platanus orientalus, Juglans regia, Acer turkestanicum* and other sp.
	Middle elevation mountains of Northern Tajikistan	Broad-leaved xeromorphic forests: *Amygdalus bucharica, Acer regelii;* Forests of *Juniperous seravshanica*
	High mountains of Western Pamir	Fragments of broad-leaved and small-leaved forests with *Juniperus semiglobosa* and *J. turkestanica*
	High mountains of Eastern Pamir	Fragments of small-leaved and microthermic forests with *Juniperus semiglobosa* and *J. turkestanica*
Turkmenistan	Mountains of Kopetdag, Great and Small Balkhan, Kugitang	*Xerophylous woodland: Juniperus turkomanica, Pistacia vera, Acer* sp.
	Accumulative plains (Karakum desert)	*Psammophylous* woodland: *Haloxylon* sp., *Salsola* sp., *Calligonum* sp.; Forests of river valleys *Populus* sp., *Salsola* sp.
Uzbekistan	Middle elevation mountains Ranges of Charkal'sky, Kuraminsky and Gissarsky	Coniferous forests of *Picea schrenkiana*; Forests of *Pistacia vera, Amygdalis bucharica* and *Juglans regia*
	Accumulative plain (Kyzylkum desert)	*Psammophylous* woodland : *Haloxylon* species, *Salsola* sp., *Calligonum sp.*; Forests in river valleys (*Populus* sp., *Salix* sp.)

Table 6.5. Areas of forests and growing stock (National Economy of the USSR, 1991).

Republics	Percentage of forest land	Forest area per habitant, hectares	Growing stock, mln. m^3
Kazakhstan	3.16	6.3	366.0
Kirghizstan	3.3	6.3	23.0
Tajikistan	1.6	0.1	6.0
Turkmenistan	12.1	3.0	14.0
Uzbekistan	11.6	0.6	11.0

Table 6.6. Industrial wood cutting and the size of forest area (National Economy of the USSR, 1991).

Republics	Industrial wood cutting in state forests, thousands ha						Forest area, thousands ha		
	1981	1983	1984	1987	1988	1989	1983	1989	Change
Kazakhstan	85.4	87.2	72.7	71.8	71.2	65.9	9353	9643	+288
Kirghizstan	0.7	-	-	-	-	-	737	729	-8
Tajikistan	0.2	36.1	31.1	32.5	32.5	32.5	458	410	-48
Turkmenistan	-	-	-	-	-	-	5121	4127	-994
Uzbekistan	8.4	10.0	10.1	10.2	4.3	5.8	1909	1302	-392

Table 6.7. The volume of wood removed from the forests of Central Asian republics, thousand m^3 (National Economy of the USSR, 1980, 1991).

Republics	1980		1985		1989	
	Total	Including industrial timber	Total	Including industrial timber	Total	Including industrial timber
Kazakhstan	2183	1674	2713	1766	2515	1926
Kirghizstan	35	14	30	7	29	7
Tajikistan	16	1	10	-	-	-
Turkmenistan	3	-	4	-	4	-
Uzbekistan	67	-	28	-	22	-
Total	**2,304**	**1,689**	**2,785**	**1,733**	**2,570**	**1,933**

Table 6.8. The volume of nuts stored in the forests of Central Asia (Isaev, 1991).

Republics	Stored up, metric tons			Target for 1990
	1975	1980	1985	
Kirghizstan	120	140	150	200
Tajikistan	500	480	520	550
Turkmenistan	9.5	15	20	25
Uzbekistan	25	27	130	330
Total	**654.5**	**662**	**820**	**1,125**

The main medical plants are the following: *Poligonum* sp., *Hypericum* sp., *Thermopsis* sp., *Glycyrrhiza glabra, Xenanthemum* sp., *Plantago mayor, Artemisia* sp., *Inula* sp., *Salvia* sp., *Matricaria* sp., *Achillea tenuifolia, Salsola richterii, Geranium pratense*. Food plants includes the following species: *Bunium persicum, Allium* sp., *Rheum maximowiczii, Ziziphora* sp. Cork of some plants is used as tanning matter or dye-stuff. To our regret, official statistics include only the total data on collection of many useful plants, without indicating the plant species (Table 6.10).

Numbers of useful wild plants were reduced during the last decades because of rapacious exploitation and cutting forests, soil contamination and agricultural development of virgin lands. Some plant species were included in "The Red Data Book" (See the Section 6.1 of this Book). At the same time the needs for these plants has increased from year to year, and the volume of stored up plants does not satisfy the demand for these resources.

Table 6.9. The volume of wild fruits stored up in the forests of Central Asia, MT (Isaev, 1991).

Republics	1975		1980		1985		Target for 1990	
	Total	Including apples	Total	Including apples	Total	Including apples	Total	Including apples
Kirghizstan	1,500	1,300	1,300	1,000	230	200	700	600
Tajikistan	1,600	400	2,000	550	2,500	2,500	2,800	900
Turkmenistan	-	-	-	-	-	-	-	-
Uzbekistan	480	100	560	120	650	130	700	140
Total	3,580	1,800	3,800	1,670	3,380	980	4,200	1,640

Table 6.10. The volume of medical and food plants stored up in the forests of Central Asia (Isaev, 1991).

Republics	Stored up, metric tons			Target for 1990
	1975	1980	1985	
Kirghizstan	0.5	8	10	12
Tajikistan	250	420	810	850
Turkmenistan	3	1	2	3
Uzbekistan	215	320	500	450
Total	468.5	749	1,022	1,064

Protective and sanitary functions of forests

We don't have historical data on the status of forests in Central Asia in the past. Only one example can be given. Bykov (1880) visited the Guzar Mountains, namely the ranges of Kuitantau and Akbash. He found that vast areas were covered with archa forests at that time. This species was met at an elevation of 6oo m and higher above sea level. Rakhmanov (1994) who visited the same areas 100 years later noted the deplorable status of archa forests which had been destroyed by illegal cuttings. The lower altitude limit of archa forests ranged from 600 m to 1,200 m above sea level. He also found that of 90,000 hectares of archa forests in the past, only 30,000 ha remained.

Cutting mountain forests resulted in severe water erosion and mud flows. On deforested mountain slopes, runoff of mud and stones is ten times greater than in the forest. Heavy rains fell in Central Asia in May – June 1992 and mud flows destroyed many houses and caused other damage. According to Safarov & Safarov (1995), land-slides in Tajikistan destroyed 6,000 houses; they claimed 140 dead people and vast areas of cotton plantations were washed away. The damage totaled 10 billion roubles. In Kirghizstan 10, 000 houses were destroyed, 40% of the cultivated land was covered with mud and stones. The country needed 2.5 billion roubles to overcome the consequences of this disaster. In Turkmenistan 220,000 ha of cotton plantations were covered with mud.The disaster had several victims.

This catastrophic flood should be a serious warning to the governments of Central Asian countries. According to published information, other environmental catastrophes had been threatening the countries of the region (Anonymous, 1997b). The question is of the Sarezskoe Lake located in the Pamir mountains. The lake appeared on topographic maps in 1911 when a land-slide blocked the flow of the Murgab river. A new lake, 79 km in width and 500 m in depth was formed. According to the preliminary estimation of scientists, an environmental catastrophe could embrace vast areas if the artificial dam would be destroyed. In this case 32,000 km^2 would embrace a part of the territories of Tajikistan, Afghanistan, Uzbekistan and Turkmenistan. In spite of this threat the government of Tajikistan, involved in civil war didn't study the problem.

All forests of Central Asia play very important roles in the life of people. Non-resource features and the benefits of forests are very difficult to express in term of money. Forests protect watershed areas and reduce surface runoff, they also play an hygienic and sanitary role.

The best forests have a volume density (crown density in desert), of 0.4 and more. As known, forest plantations protect agricultural crops from drought and dust storms, from sand drifts and water erosion.

According to published data (Isaev, 1991), Central Asian forests fulfill sanitary hygienic and prophylactic functions only on an area of 29.5 thousand hectares. For example, 1 ha of archa forests secretes 5 metric tons of phyton cides per 24 hours! Several other tree species, like *Juglans regia, Acer nrgundo* and *Aesculus hippocastanum* have a high ability to produce phytoncids. Other species, like saxaul, *Salsola richterii, Tamarix* species and some others have very low phytoncids activity.

Forests can also accumulate dust and clear the air, especially in the layer of 0 – 40 m a bove the ground. Plantations of deciduous trees can filter out 50 – 70 metric tons of dust per hectare per year. From trees of Central Asia, the following species are the most resistant against dust and gases: *Sophora* sp., *Morus alba, Betula turkestanica* and *Gleditschia triacanthos* (Isaev, 1991). The influence of forests on human health is especially revealed from June to September when the air temperature is high. The comfort produced by forest is so great that sanatoria, pioneer (scout) camps and tourist centers are usually located in forests.

Degradation of forests during the Soviet period

A special scientific expedition of the USSR State Department of Forestry conducted a study of forests in Central Asia. Members of this expedition organized field missions in all republics of Central Asia. The results of this study were published by Sukhikh (1991). The author concluded the following:

1. Productivity of forests during the last 50 years decreased. Especially diminished were the areas of archa, walnut-tree (*Juglans regia*) and tugai forests. Tugai forests covered 52.9 thousand ha, their growing stock had changed from 120 – 200 m^3/ha to 40– 90 m^3/ha. Valuable archa forests covered 566.9 thousand ha and 85% of this area had poor regeneration.

2. As a rule natural regeneration of forests is poor. Artificial regeneration didn't give good results. In 1951–1990 work on artificial forest regeneration was conducted on an area of 5.7 mln. ha. From this area only 1.7 mln. ha were conserved and registered as productive forests.

3. The area of fir and spruce forests was also reduced. 47% of them were in temporary use of collective farms as grazing lands. Natural regeneration was poor on 85% of their area.

4. According to recommendations of scientists, the sub-region needed 8.8 mln. ha of protective forests and forest belts. But in 1992 only 910.2 thousand ha of such forests were registered. They included 226.8 thousand ha of erosion protection stands, 497 thousand ha of pasture protection stands and 186.1 thousand ha of field protection belts. The rest belonged to other categories of forests.

5. 18 mln. ha of protective forests were planted during 1968 – 1990, including 1.2 mln. ha of pasture protective forests, 0.4 mln. ha of erosion protective bel ts and 0.2 mln. ha of field shelter belts. A considerable part of these stands perished, the rest were in deplorable state. Protection forest stands were not fenced, they were damaged by livestock, especially young stands.

6. In Uzbekistan and Kirghizstan special measures were taken to increase the production of walnut forests. In Turkmenistan the area of *Pistacia vera* stands increased, especially in Kushka forest district.

Sukhikh (1991) also indicated many causes of forest degradation in Central Asia, in particular: animal grazing, unwarranted cutting, poor scientific justification of measures on forest regeneration, low salary of forest personnel, poor equipment of forest enterprises, etc. According to published map (Kharin, Kalenov & Kiril'-tseva et al., 1993), degradation of vegetation in Kirghizstan was registered on an area of 47,245 km^2. The same map gave information on the degradation of forests in Tajikistan. Forests were degraded on an area of 71,6333 km^2, of which 81,9% fell on moderately degraded land and the rest – on slightly degraded land. These figures on Tajikistan needed correction because civil war and economic collapse aggravated the destructive ecological process.

We have limited information on unwarranted wood cutting in Central Asia. In many cases, rangers themselves violated the Forest Code by selling firewood. For example, M. Redzhepov, forester of Kopetdag district in Turkmen SSR, ploughed up 700 hectares of saxaul forest. He wanted to use this land for cultivation of agricultural crops. But the forest administration of this district for a long time didn't want to institute proceedings against him. After great delay he was sentenced and paid the damage (Ovcharenko & Chernenko, 1988).

Many cases of unwarranted cutting were registered in state forests of Turkmenistan. 603 cases of violation of the Forest Code were registered in 1991,905 cases in 1992. In the first case 458.8 m^3 of wood were cut, in the second case 780.5 m^3 of wood (National Action Program .., 1996). But it was only "a small drop in the ocean of unwarranted cutting"... The low salary of rangers and other forest personnel in the USSR stimulated violation of the Forest Code, especially in regions poor in wood. Unwarranted selling of timber and firewood was an illegal source of income for many persons working in forestry.

Reduction of the area of woodlands in the deserts of Central Asia has been a special alarm to scientists. Stands of *Haloxylon* species and other shrubs have been subjects of severe exploitation for centuries. The Tsarist administration couldn't solve the problem. During the Soviet period the problem of conservation of desert woodlands was ignored. In all Central Asian republics, desert woodlands formally were under the administration of the departments of forestry. But by special decrees of governments of these republics, desert woodlands were given to departments of agriculture for long-term use as grazing lands. So, one department was responsible for conservation, the second department was responsible for exploitation of these lands. As said, "too many cooks spoil the broth".

Use of machinery during the "age of technical progress" did enormous damage to saxaul stands. Many years ago, people cut or better to say broke off branches of this shrub not disturbing the roots. In many cases that even stimulated the regeneration of saxaul and other shrubs. Poachers of "the age of technical progress" rooted out saxaul with special wire ropes dragged by tractors or by lorries. That was a real "remaking of nature" because the absolute desert appeared in these places!...By rough estimate, some 5 mln. hectares of saxaul stands have been destroyed during the last 50 years (Kalenov, 1992).

6.3. Degradation of desert rangelands

The main causes of desert rangeland degradation

The causes of land degradation in Central Asia are given below, in percentage of area affected by desertification (Kharin, Kalenov & Kiril'tseva et al.,1993):

Vegetation cutting	28
Overgrazing	29
Excessive irrigation	9
Construction of roads and canals	8
Hay making	3
Undergrazing	3
Water filtration	2
Other	18
Total	100

The category "other" includes fires, mine working, movement of transportation vehicles and some other non-identified causes. Degradation of the vegetative cover is the main type of desertification in Central Asia. Usually, in the initial stage of human interference in the desert environment, vegetation is first of all damaged by transportation vehicles or is cut for fuel.

Change of dominants is a very important criterion by assessment of rangeland degradation. According to Kalenov (1992), 60.5% of the area in the Karakum desert occupied climax or slightly changed plant associations, 22.9% fell to plant associations in which dominants were partially changed, 3.5% belonged to ephemeral (short-living) associations, the rest belonged to other land cover types (oases, salina, moving sand dunes, etc.).

A system of free grazing dominates in the range management in Central Asia. Degradation of desert rangelands is a result of rapacious exploitation of plant resources. This process was aggravated after 1991 when changes in the political system and collapse of the economy aggravated the ecological situation. The degree of rangeland degradation depends upon the composition of livestock and intensity of grazing. Sheep graze herbaceous plants, camels mainly pluck shrubs and semishrubs, goats can do the most serious harm to all types of vegetation. Animal pressure is the main factor affecting degradation of rangelands. Kharin, Nikolaev, Sorokina & Kiril'tseva (1989) found that the correlation between the area affected by desertification and animal pressure can be proved statistically. From 39 administrative districts in Turkmenistan studied, 16 had high animal pressure, 13 had moderate pressure and only 13 had slight animal pressure. Regression equations for Turkmenistan are given in Table 6.11.

Table 6.11. Regression equations for estimation of percentage of area affected by desertification (Kharin, Nikolaev, Sorokina & Kiril'tseva, 1989).

Administrative regions	Number of counties	Equations	r
Krasnovodsk	6	$y = 30.095 + 6.198x$	0.674
Ashkhabad	8	$y = 25.431 + 0.341x$	0.808
Mary	8	$y = 30.851 + 0.240x$	0.364
Chardzhou	10	$y = 26.044 + 0.179x$	0.485
Tashauz	7	$y = 17.831 + 0.514x$	0.954

Note: y – percentage of area with moderate and severe desertification; x – actual animal pressure, in percentage of potential pressure; r – correlation coefficient.

Vegetation degradation around watering points

Concentration of livestock around watering points has led to the formation of desertification spots, 2 to 10 km in diameter. Each desertification spot has a circular structure, namely:

1. So called "tyrlo" which is located in the center of spot, around the well. This place is devoid of vegetation because livestock is concentrated here for watering and spending the night. Koshara (shed), yurta or saman houses are also located here,
2. The ring of moving sand dunes originated because of intensive movement of flocks,
3. The ring of eroded sand dunes, partially covered with non palatable plants. The flocks move here towards the grazing lands,
4. The ring of movement of flocks where vegetation is severely degraded. A zone of dispersed grazing is located beyond the 4th ring.

Kharin & Kalenov (1978) identified these spots by space photos. The size of desertification spots in two regions of Turkmenistan is shown in Table 6.12.

Regional features of rangeland degradation

The process of vegetation degradation under the impact of grazing has the following common feartures:

1. Palatable plants are replaced by less palatable and unpalatable species,
2. The soil surface is disturbed by grazing, so plant species growing on loose soil are increased in number,

Table 6. 12. The size of desertification spots around desert wells, km (Kharin & Kalenov, 1978).

Regions	n	d	S	<1	1 – 2	2 – 3	3 – 4	4 – 5	5 – 6	6
Interfluve Tedzhen-Murgab	66	2.21	0.83	6	22	18	5	1	3	1
Southeast Karakum	94	2.16	1.26	1	56	28	6	2	-	2

Note: n – number of spots; d – mean diameter, km; s – standard deviation, km.

3. Long overgrazing is results in the change of plant communities, and climax communities are replaced by anthropogenic vegetation,
4. Production of biomass is reduced and quality of forage is becoming worse,
5. Because of reduction of biomass and deterioration of forage quality lands suitable for the whole year round grazing are transformed to seasonal grazing lands.

Bereznev (1995) found that in several provinces of Kazakhstan forage production of desert rangelands was exhausted even by the sixties of the last century. In spite of that, the number of livestock increased by 40% during the last 25 years. At present, 25% of rangelands provided with water sources is subject to degradation. In general the livestock is provided with pasturage by 51.8%. At the same time several remote areas in Kazakhstan have insufficient animal pressure because of the shortage of water (Abdraimov, 1990). A new report on degradation of rangelands in Kazakhstan was published by Baitulin & Bekturova (1997) and given in Table 6.13.

In Turkmenistan the process of rangeland degradation was especially intensified during the last decade. During this period 10,000 ha of rangelands lost their productivity annually. Areas of degraded rangelands in Turkmenistan are given in Table 6.14 (National Action Program...,1996), statistics on the 1 January 1993.

Table 6.13. Desertification of grazing lands in Kazakhstan, in mln. ha (Baitulin & Bekturova, 1997).

Zones	Area of rangelands	Desertified	Severely desertified
Forest and steppe	32.3	12.8	3.6
Central hill area	29.2	9.0	0.8
Semidesert	21.0	8.4	2.4
Desert	73.7	44.6	17.2
Mountains	9.4	4.5	1,0

Table 6.14. Area of degraded rangelands in Turkmenistan, km^2 (National Action Program.., 1996).

Velayat	Slight desertification	Moderate desertification	Severe desertification	Total
Akhal	42433	42401	2412	86246
Balkan	43116	43583	6470	93169
Tashauz	20336	39905	903	61144
Lebap	44428	26766	2007	73201
Mary	46982	22153	5461	74506
Total	**196295**	**174808**	**17253**	**388356**
%	50.5	45.0	4.5	100

Information on degradation of the vegetative cover in Uzbekistan was published by Kharin, Kalenov & Kiril'tseva et al. (1993). From 214,480 km^2 of desert rangelands and forests, 58 % were slightly degraded, 37% were moderately degraded and 5% were severely degraded.

In conclusion of this section, we want to give information about hay harvesting in the former Soviet Republics of Central Asia. It was an additional source of forage used in case of emergency. Hay was harvested and stored in forests, on rangelands and on meadows, especially during wet years. It was stored up by kolkhoz, sovkhoz and forest farms (Table 6.15). A part of hay stored was given to a special emergency reserve created in each republic. This system was destroyed after 1991.

Table 6.15. Hay reserve harvested in four republics of Central Asia (Isaev, 1991).

Republics	Volume of hay stored, thousands metric tons			Target for 1990
	1975	1980	1985	
Kirghizstan	11.8	15.3	13.0	14.0
Tajikistan	6.2	8.0	9.3	9.8
Turkmenistan	-	-	7.0	7.5
Uzbekistan	3.8	7.8	11.8	14.1
Total	21.8	31.1	41.6	45.4

6.4. Soil degradation

Environmental changes in the Aral Sea zone

Drying up the Aral Sea was caused by withdrawal of water of the Amudarya and the Syrdarya for irrigation. By the assessment of Kharin, Kalenov & Kiril'tseva et al. (1993), salinization of soils due to the drop of the Aral Sea level was registered on an area of 49,197 km^2. A new anthropogenic desert was formed in the vast area. That resulted in development of desertification process, in particular: blowing out and accumulation of salt and dust, degradation of tugai vegetation, fall of underground water table, change of sediments in river water, increase of salinity of river water, etc. Razakov (1990) found that 15 – 75 mln. metric tons of salts were blown of the dried bottom of the Aral Sea. These salts were tran-sfered by the wind at great distances. Accumulation of these salts was registered in the region of low Volga.

In the former Amudarya delta the process of environmental changes has been stabilized during the last 10–15 years. The size of solonchak area has been increased by 7%. Environmental degradation in the former bay of Adzhibai is more intensive (Krutov, 1995).

Considerable environmental changes occured in the Syrdarya delta. They were caused by drying up the river channels and numerous lakes and by sinking of the underground water table. Physical and chemical properties of soils were drastically changed. For example, the content of humus in the upper 50 cm soil layer decreased by 13 - 27 metric tons per ha in meadow soils and by 33 - 35 metric tons per hectare in swampy soils. By 1990 the following changes occured in the area under consideration:

1. 114 thousand hectares of alluvial-meadow soils were transformed to solonchak,
2. 532 thousand hectares of swamp and meadow-swampy soils became dry,
3. 31 thousand hectares of other soils degraded and transformed to waste land.

In general, about 732 thousand ha of soils have lost their productivity, and they can't be used as rangelands or agricultural lands.

Wind erosion

Wind erosion mainly occurs in sandy desert. Many scientista publish information on wind erosion and give different figures about affected areas. That can be explained by the different methodological approaches. In the assessment of wind erosion, the following three points should be taken into consideration:

1. In the Karakum desert natural erosion (deflation) is observed on 15% of sands grown by vegetation. These eroded areas include blow-outs, single barkhans or groups of barkhans. They usually are met on the upper parts of sand ridges. Their location can be mapped only by large scale aerial photos because the size of these erosion spots is very small.
2. Tracts of moving sand dunes of natural origin should be excluded from assessment of human-induced land degradation. There are several tracts of such sands, their origin is not connected with human activity.
3. Tracts of moving sands of anthropogenic origin should be identified by the assessment of desertification. These sands are confined to areas of active human interference in desert environment like roads, canals, construction sites, etc.

In Kazakhstan wind erosion affected 20.5 mln. hectares of ploughland and 25 mln. ha of rangelands. Technogenic desertification occured on the area of 181.3 thousand ha. These areas included gas pipelines, mining sites and military testing sites of the former USSR (Baitulin & Bekturova, 1997).

Water erosion

Water erosion is developed on mountain slopes and on piedmont plains of Central Asia. According to the map (Kharin, Kalenov & Kiril'tseva et al., 1993) water erosion was registered on the area of 82,578 km². Slightly eroded areas dominate (64%), the rest are moderately degraded. In the drylands of the sub-region, water erosion is stimulated by heavy rains and by melted snow. Adverse human activities accelerate the process of water erosion. Inherent risk of water erosion includes mechanical composition of soil, slopes, stones on the earth surface and precipitation rate.

Nurberdyev (1982) studied soil erosion in Turkmenistan. He described the process of formation of erosion landforms on the Kopetdag piedmont plain. On the initial stages the surface runoff forms rills which are then transformed to wash-outs, and the last become ravines with many branches. He identified three belts of erosion on mountain slopes of the Kopetdag:

1. Heavy erosion on upper parts of slope with thin humus horizon, 10 – 15 cm,
2. Moderate erosion of middle slopes with humus horizon 15 – 20 cm in depth,
3. Slight erosion of alluvial cones and foothills with humus horizon 25 – 35 cm.

The same author also identified water erosion on slopes where dry agriculture (bogara) is practiced. Wheat, barley and local varieties of peas are cultivated in such places. Development of erosion depends upon slopes, exposure and technique of tillage. In spite of the recommendations of scientists ploughing along slopes has been used everywhere. That stimulates rill erosion and washing out the upper soil horizons. Loss of agricultural production could reach 25% by slight erosion, 25 – 50% by moderate erosion and more than 50% by severe erosion. According to Baitulin & Bekturova (1997) water erosion in Kazakhstan affected 20.5 mln. ha, among which 5.2 mln. ha were severely affected. Annually in this country 60 mln. metric tons of soil material are washed away from plough lands. The erosion on irrigated fields were registered everywhere on the area of 1.8 mln. ha. Soil dehumification was common on11.2 mln.ha of virgin lands in the steppe zone.

Waterlogging rangelands

In all oases of Central Asia, over-irrigation and leaching of saline soils are the causes of the accumulation of plenty of used water. This superfluous water gives a lot of troubles to local administrators and planners. Great amount of this water is conducted from oases to the desert. This water floods rangelands and forms many new streams and lakes. As a result, productive desert rangelands are put out of operation because of the following reasons:

1. Palatable forage plants are replaced by hydrophytes and pheatophytes (*Tamarix sp., Phragmitis australia* and other plants),
2. New streams, bogs and lakes hinder free movement of flocks,
3. Not only affected lands but also surrounding areas are excluded from exploitation because of the reason indicted in the point 2.

According to the map (Kharin, Kalenov & Kiril'tseva et al., 1993), desert rangelands are affected by waterlogging in an area of 34,504 km². At present, melioration of these lands is practically impossible because of the lack of money.

Salinization of irrigated farmlands

Grandiouse works on construction of irrigation canals and reclamation of new lands have led to changes in the physical environment in Central Asia. Some areas are transformed to the zones of ecological catastrophe, like the Tashauz oasis. In other areas, newly reclaimed lands lost their productivity in 8 – 10 years. The ecological changes caused by irrigation have been described by many scientists (Dukhovnyi, 1980; Zonn, Nikolaev & Orlovsky, 1981; Kovda, 1977, 1989). Table 6.16 gives the classification of the ecological changes by irrigation (Dukhovnyi, 1980).

The author of this table considers as stable those changes which always take place by irrigation. Non stable changes do not always take place, they could be observed or not. From 8,672 mln. ha of irrigated farmlands in Central Asia about 90% is affected by soil salinity, from which 10% needs melioration of the whole area (Kharin, 1996). Cotton is the dominant crop on irrigated soils of Central Asia. So the loss of yield of this crop is the main criteria of assessment. Salt accumulation also hinders the development of roots

Table 6.16. Change of environmental factors under the impact of irrigation (Dukhovnyi, 1980).

Type of changes	Changing factors	Positive effect	Negative effect
Stable, not controlled	Milding the climate	+	
	Change of natural vegetation	+	
	Change of microorganisms	+	
	Change of wildlife	+	
	Lowering the force of wind	+	
	Change of river flow	+	
	Change of cultural vegetation	+	
	Lowering non productive evaporation	+	
Not stable, controlled	Change of ground water level	+	-
	Change of humudity in aeration zone	+	-
	Change of ground water mineralization	+	-
	Change of river water mineralization	+	-
	Change of physical property of soil	+	-
	Change of salt content in the aeration zone	+	-
	Erosion of slopes		-

of cotton and fruit-trees. Crops without main roots, like wheat and barley suffer less.

Poor agrotechnique, insufficient drainage network and nonobservance of crop rotation are the main factors stimulating secondary salinization of irrigated farmlands. During the last decades this process was intensified on account of the use of mineralized water with salt content 2 g/l and more. In the initial stage soil salinization appeared on small spots, in the second stage these spots are joined in big patches, in the third stage the whole area is affected. In the southern part of the Aral Sea zone, 20.3% of irrigated soils is severely affected by salinization, 20% is moderately affected and 59.6% is slightly affected (Rafikov, 1984). In Khorezm province of Uzbekistan slightly salinized soils dominate. Severely salinized soils are only met in patches. Observance of agrotechnique and soil leaching hinder the worsening of soil conditions (Tursunov, 1981).

In the Zeravshan delta, salinization of irrigated soils is observed in small patches (Gafurov & Abdullaev, 1982). In Karshi oasis mineralization of ground water exceeds 10 g of dry salt per litre. Soils are moderately affected by salinization. In Sharabad oasis the yield of cotton totals 2.5 – 2.8 metric tons per ha, that testifies to slight and moderate salinization.

Irrigated soils of Turkmenistan are also affected by salinization (National Action Program.., 1996). 47% of irrigated farmlands have the depth of ground water less than 2.0 m. Use of mineralized water for irrigation, absence of crop rotation and application of large amount of fertilizers stimulate soil degradation. As known, optimal quantity of fertilizers stimulates the productivity of agricultural crops and soil fertility. But if a large amount of fertilizers is applied, the surplus of nitrogen can depress soil biota. By degree of soil salinization, irrigated farmlands of Turkmenistan are divided into the following classes (in percent): non-saline – 4.6, slightly saline – 28.5, moderately saline – 53.8, severely saline – 13.1.

Especially severe soil salinization was registered in a Khauzkhan tract of newly developed lands. Striving for high cotton production by neglecting agricultural technique has led to severe desertification. The depth of ground water totals 2.0 m and less, mineralization of drainage water reaches 40 g of dry salt per liter. As a result, the yield of cotton has decreased to 1.0 metric ton per ha. In 8 – 10 years the tract of productive soils was put out of operation. In the Tashauz oasis of Turkmenistan, degradation of irrigated soils during 10 years (1979 – 1988) was described by scientists (Table 6.17).

During this short period of time, the physical properties of soils and the productivity of cotton plantations have fallen.

Salinization of irrigated soils is a serious problem in Kazakhstan. The soils of secondary salinization here occupy 376.7 thousand ha, 20% of all irrigated soils. About 0.5 mln. ha of irrigated soils are not used because of salinization and the problem with irrigation systems. The size of the area which is not used increased five times after 1990 (Baitulin & Bekturova, 1997).

Table 6.17. Characteristics of soil/water situation in the Tashauz oasis (Korzh, 1990).

Years	Total area, thousand ha	Length of drainage network, m/ha	Water given for irrigation, thousand m^3/ha	Water conducted from oasis, thousand m^3/ha	Salt content in drainage water, g/l	Depth of ground water, m	Salt content in irrigation water, g/l	Yield of cotton, metric tons/ha
1979	211.0	27.8	27.3	5.4	2.93	1.84	0.73	3.14
1982	243.5	27.5	22.1	4.0	4.32	1.97	0.96	2.68
1985	273.0	26.5	25.2	4.3	3.88	1.85	1.01	2.50
1988	320.7	26.4	19.2	5.0	4.74	1.86	0.98	2.32

6.5. Change of desert environment along the Karakum Canal bordering area

Construction of the Karakum Canal caused great direct and indirect environment changes in vast areas of Central Asia. These changes were first of all registered in the comparatively narrow strip of land bordering the canal itself. During the construction of the canal, new tracts of moving sand dunes appeared in sandy deserts because of the influence of machinery, vegetative cover being also destroyed. Our experimental area was located in southeast Karakum, in the vicinity of the small villages Kara Belend and Nadyr Belend.

The experimental area, 20/40 km in size included sandy desert with various landforms. Primitive sandy soils were underlaid by a thick layer of alluvial sands of the Pra Amudarya (ancient Amudarya) river. The traditional approach in classification of moving sands in the former USSR was based on assessment of topography, water regime and nutritive substances of soil which were necessary for plants in case of phytomelioration. A meliorative classification of sands in the area of consideration is given in Table 6.18 (Kharin, Nurberdyev & Serkhenov, 1969).

The description of these sand types is given below.

1. Moving sands appeared close to the canal bed; their origin is connected with ground excavation works during the canal construction. These sands are also known as technogenic sands because they were formed during the digging of the canal. Barkhans wre 1 m in height; in some places the dam was stabilized by cultivation of *Salsola paletzkiana* and other plants. Natural plant cover changed under the impact of man. *Haloxylon aphyllum,* which had formed climax communities before the interference of man, was replaced by *Haloxylon persicum, Salsola richterii* and *Calligonum setosum (Figure 6.1).*

Barkhan sand dunes have a favorable water regime for plant growth. The content of water in the root zone reaches 2 – 5% during the whole year. Only the upper layer above 30 cm dries during summer. The next strip of sands, by move-

Table 6.18. Characteristica of sand types in the canal bordering area (Kharin, Nurberdyev & Serkhenov, 1969, modified).

N	Meliorative types	Geomorphology	Depth of ground water, m	Number of shrubs/ha	Crown density of shrubs, %	Number of *Carex physodes* plants/m^2
1	Moving and slightly grown sands	Sand ridges and barkhans	1 – 5	10 – 50	2 – 10	50 – 100 (none on barkhans)
2	Sands surrounding filtration lakes	Sand ridges and basins	1 – 3 m near lakes, 10 m on elevations	130 – 180	20 – 40	200 – 300
3	Sands grown by vegetation	Sand ridges and honecomb sands	5 – 10	180 – 250	30 – 40	160 – 250
4	Sands of complex topography grown by vegetation	Sand riges, basins and small hills	>10	150 – 180	30 – 50	150 – 200
5	Sands surrounding desert wells, slightly grown by vegetation	Slightly undulated sands	>10	100 – 150	20 – 30	50 – 150
6	Slightly undulated Gypsum sands	Plain topography, small hillocks	>10	50 – 100	10 – 20	None

Figure 6.1. Crumbling banks of the Karakum canal. Bare roots of desert shrubs are seen on foreround. Photo by A. Prishchepa.

ment from the canal, is slightly covered by vegetation. The content of salt here is small, less than 1%, in which the chlorine ion is 0.02% and the sulphate ion 0.01 – 0.03%. The content of humus is very low (0.10 – 0.15%), carbonates vary within 2.5 – 3.8%.

2. Sand surrounding filtration lakes. According to site conditions, these sands can be divided into two sub-types:

a. Narrow strips, 3 – 10 m in width located close to lakes with vegetation of tugai type (*Tamarix* sp., *Phragmites communis, Carelinia caspica*). Because of high evaporation the upper soil horizon is strongly salinized (1.62 – 2.60%).

b. Ridgy sands with small basins form the second belt around filtration lakes. These sands have been covered by vegetation (*Haloxylon persicum, Calligonum* sp., *Salsola richterii, Carex physodes,* etc.).

3. Sands covered by vegetation. They are located to the north of the strip of filtration lakes. Sand basins have a depth of 5 –7 m and 30 – 50 m in width. Psammophylous vegetation consisting of *Haloxylon persicum, Calligonum cetosum* and *Carex physodes* dominates here. The influence of man is very limited here, mainly grazing of sheep and wood cutting for fuel can be distinguished on these sands.

4. Sands of complex topography covered by vegetation. These sands have much in common with the 3rd type, but topography here is more desiccated: sand basins are 8 – 15 m in depth and 70 – 130 m in width. Climax communities of *Haloxylon aphyllum* + *Carex physodes* dominate here. These sands don't need any meliorative works, only measures on conservation of plant cover can be recommended. Black saxaul is usually cut for fuel.

The sands of types 4 and 5 also don't need any melioration. Sands surrounding desert wells had been formed on account of the concentration of flocks near watering points. But after canal construction, these wells lost their importance because herdsmen prefer to water animals in the canal.

There are different approaches in the phytomelioration classification of sands. For example, Kharin & Petrov (1975) noted many physical factors which should be taken into consideration, namely: parent material, depth of ground water, topography, content of salt in soil horizons, types of sand dune migration, density of vegetation, velocity of sand movement, etc.

For assessment of the phytomelioration features of sands, we need first of all information on the degree of sand stabilization by vegetation. Many desert scientists use as an indicator of sand surface stabilization the density of *Carex physodes* (local name "ilak"). Ilak is an ephemeroid with an extensive root system and rhysomes (Figure 6.2).

Above that, this plant is a valuable forage plant. Different authors give different recommendations on this topic. For example, Nechaeva, Prikhod'ko, Bashkatova & Kiyanova (1959) identified the following three categories of sands:

1. Stable sands (sands covered by vegetation) with the total plant cover (shrubs and semi-shrubs) 30 – 50%, including the ilak coverage 15 – 20%,
2. Sands moderately covered by vegetation with the total plant cover 15 - 20%, including 3% of ilak,
3. Moving sands (barkhan sands) with 2 – 7% of shrubs; ilak in these places is absent.

Lavrov (1964) had the same approach but his criteria are a little different. In his opinion, ilak is also the indicator of sand stabilization, but his criteria are as follows: stable sands (200 – 300 plants of ilak per m^2), sands moderately covered by vegetation (100 – 200 plants of ilak per m^2), and moving sands which have less than 100 plants of ilak per m^2.

A special classification was developed for assessment of sand stability in sandy desert for practical application. As Technical rules... (1963) propose, the percentage of sand surface, stabilized by vegetation, is used for classification of sands by the construction of roads in sandy deserts. Three categories of sands are identified: stable sands (more than 35% is stable, that is covered by vegetation), semi-stable sands (35 – 15% is stable), moving sands (less than 15% is stable).

These publications have shown that the most suitable classification of sands on the degree of stability included three categories: stable sands,

*Figure 6.2. Dense cover of ilak (*Carex physodes*) which stabolize the sand surface. Photo by A. Prishchepa.*

semi-stable sands and moving sands. From these categories the first and the third are the most definite, the second category is less clear.

Coming back to our experimental area, we can say that the origin of all moving sands here is connected with human activity – construction of the Karakum Canal. We used large scale aerial photos for mapping these sands (Figure 6.3).

The main tracts of barkhan sands were confined to a 1 km strip located close to the canal bed where 30% of the total area had been destabilized during canal construction. Especially dangerous for the canal exploitation were barkhans located on the banks of canal. Sand from barkhans moved into canals and filled up their beds. So, for normal functioning of the canal, special excavation works should be conducted continuously. These areas also needed special measures on sand stabilization. They included mechanical sand stabilization combined with phytomelioration. On the rest of the experimental area, moving sands occupied less than 10% of the land.

Protective measures for conservation of saxaul stands and limited grazing can keep the ecological equilibrium in this area.

6.6. Summary and conclusion

The Red Data Book of the USSR includes 92 species of flora in Central Asia. From them, the rare plants number 46 species and 46 species have decreased in number.

The drop of the Aral Sea level has changed the ecological situation in the area of 49.2 thousand km^2, representing 12.3% of the total area of Central Asia. These areas were changed into waste land.

7 – 10% of desert rangelands lose their productivity annually. The status of desert rangelands is especially severe in Kazakhstan, where forage production was exhausted even by the 1870s. Overgrazing, shortage of water resources and change of political system are the main reasons for this process.

Figure 6.3. Large scale photos of the Karakum canal bordering area. Filtration lakes were formed because of water leakage from the canal. Shrub vegetation was cut off while construction work of the canal. Only a small number of shrubs is seen, they are pictured by black dots.

Waterlogging desert rangelands on the oasis borders was registered in the area of 34.5 thousand km^2, 8.6% of the total area of Central Asia. The movement of flocks is very difficult in these areas because of many swamps and wet places.

Scientific investigations have proved, that the area of saxaul forests is reduced 10 – 12% annually . As to mountain forests, the area of low productive stands is increased from year to year because of unwarranted cutting and the loss of dominants to less valuable tree species. The area of *Pistacia vera* stands in Turkmenistan has been increased during the last 30 years. Severe salinized soils of the irrigation zone occupy 13 – 20% of irrigated lands, and 2 – 3% lose their productivity annually.

Severe wind erosion has affected vast areas in Kazakhstan, including 20.5 mln. hectares of plough land and 25 mln hectares of rangelands. Wind erosion in Turkmenistan occurs on 1.7% and in Uzbekistan 1.1% of the total areas of these countries.

CHAPTER 7

CHANGE OF WILDLIFE

We wolves are a free people!
R. Kipling – The jungle book.

7.1. Animals from "The Red Data Book"

Fauna of the region numbers 172 species of animals, 496 species of birds, 86 species of reptiles, 11 species of amphibians, 150 species of fishes and 10,000 species of invertebrates (Babaev & Kharin, 1994 b). In the process of land degradation, wildlife is changed. People exterminate animals directly by shooting, other animals perish because of destruction of their habitats. Development of new lands, cutting vege- tation and direct human pressure also lead to the change of wildlife. Conservation of vegetation will stimulate conservation of habitat (Figure 7.1).

Rustamov (1991) also found that land degradation could lead to the loss of biodivrsity at all levels. For example, after drying up the Aral Sea local flora perished because of the change of salinity of sea water from 10 to 30 g/l. In the Aral Sea zone biodiversity of mammals and birds were reduced 2-3 times.

Construction of the Karakum Canal has changed the environment. Water is a mighty factor which could change the desert wildlife population. Many animals populated new areas in the canal zone. Wild cat, fox, musteline, caracal, korsak, wild boar and other beasts came to the canal zone. The jackal became the most numerous animal in this zone. Jackals penetrated into the heart of the Karakum along 200 km of the canals, filtration lakes and other waterways. Rustamov (1991) considered as a positive factor in conservation of biodiversity the increase in number of waterfowl. 600 – 700 thousand geese, ducks and other waterfowl come to winter in the zone of the Karakum Canal.

Figure 7.1. The aerie of eagle on the tree of Calligonum arborescence*, the Kyzylkum desert, Uzbekistan. Photo by A. Prishchepa.*

Central Asian tigers have disappeared. The last tiger was shot in the region in the 1960s. Tigers were very plentiful in the valley of the Amudarya in the beginning of the century. As Rickmers (1913) said :"Russians always told me that the Duab tiger is larger and stronger than his Indian brother, but that may be a patriotic exaggeration".

Kurbanov (1985) emphasized the importance of other human activities in the destruction of wildlife. According to his observations, many animals are killed by transportation vehicles. Snakes, rodents, birds, hedgehogs and other animals died under the wheels of cars.

Degradation of tugai vegetation in the Syrdarya delta, as Mukasheva (1992) found, resulted in wildlife changes. Wild boar disappeared, mask rat, pheasant and many game birds were reduced in number. Food fishes practically disappeared in lakes and river channels. Some species of animals became so rare that they needed special measures for their conservation. The list of rare and endangered species is given in Table 7.1

Table 7.1. Rare and endangered species of fauna (The Red Data Book of the USSR, 1978).
(Note : 1 - Kazakhstan, 2 - Kirghizstan, 3 - Tajikistan, 4 - Turkmenistan, 5 - Uzbekistan)

Species	**Countries** 1 2 3 4 5	**Limiting factors**
Species populations of which are probably disappeared		
Acinonyx jubatus	+ - - + +	Reducing in number gaselle and other wild ungulates
Pantera tigris	+ - + + +	Hunting, destruction of tugai vegetation
Picus squamatus	- - - + -	Destruction of tugai vegetation
Ursus arcatos syriacus	- - - + -	Poaching, destruction of habitat, construction works
Endangered species		
Amorpha philerema	+ - + + +	Cutting *Populus* trees
Aquila chrysactos	+ + + + +	Destruction of habitat, lack of food
Aquila heliaca	- + - + -	Destruction of habitat, perish on electric power lines
Capra aergargus	- - - + -	Neglecting conservation measures in reserves
Ciconia nigra	+ + + + +	Reduction of grazing areas, cutting wood
Calasoma sycophanta	+ + + + -	Cutting wood, application of chemicals in agriculture
Catocala optima	+ - + + -	Destruction of tugai vegetation
Felis caracal	- - - + +	Decreasing in number hares, competition with wolf and fox
Francolinus franc*olinus*	- - - + -	Cold snowy winters, poaching, hay making
Gasella subguttosa	+ + + + +	Development of desert, livestock, grazing, snowy win ters
Hierodula tenuidentata	- + + + +	Cutting wood, spraying trees and shrubs with chemicals
Hyaena hyaena	- - + + +	Hunting, trapping, reduction in number of unulates

Species	Countries 1 2 3 4 5	Limiting factors
Lutra lutra scistanica	+ - + + +	Poaching, cutting wood, pollution of environment
Paragluphisia oxiana	+ - - + +	Cutting wood, fires
Rethera komarovi	- - + + -	Grazing, hay making
Rophitoides canus	+ + + + +	Application of pesticides
Scolia maculata	- - + + -	Unidentified
Varanus griseus	+ - + + +	Change of habitat, perish on motor roads
Xylocopa valga	+ + + + +	Plowing up land, cutting trees
Xylocopa violaceae	- - - + -	Unidentified
Zygaena turcmenica	+ - + + +	Land reclamation, cutting wood, vivestock grazing
<u>Rare species</u>		
Acosmerix naga	- - + - -	Cutting wood, construction works
Allolobophora (Svetlovia) ophiomorpha	+ - - - -	Cutting trees , grazing
Allolobophora *(Svetlovia)* umbrophita	+ - - - -	Cutting trees, grazing, use of chemicals in agriculture
Anthia mannerheimi	- - - + -	Land reclamation, change of habitat
Argynnis alexandra	- - - + -	Human activity
Catocala lesbia	- - - + -	Specific requirements of species to climatic conditions
Cervus elaphus *bactri*	+ - + + +	Destruction of tugai vegetation, fires
Cheimoptena *pennigera*	- - - + -	Cutting wood, grazing
Ciraetus gallicus	+ + + + +	Decreasing in number reptiles and amphibias, cutting wood
Dolbinopsis grisca	- - + - -	Cutting wood, recreation activity of people
Falco cherryg	+ + + + +	Catching, destruction of nests, reducing in number rodents
Hipocolius ampelonus	- - - + -	Destruction of tugai vegetation
Hipparchia fatua	- - - + -	Deforestation, economic development
Lynx lynx isabellinus	+ + + - -	Decreasing in number hares and deers
Mellivora capensis	- - - + -	Not studied
Pandion haliaetus	+ + - + -	Cutting wood, pollution of water bodies
Passer simplex	- - - + -	Economic utilization
Unica unica	- - + + -	Reducing in number undulates, poaching, animal grazing
Ursus arctos isabellinus	+ + + + -	Development of mountain areas
Utentheisa pulchella	+ + + + +	Cold winters, ploughing land, grazing, construction works
Vormela peregusna	- - - + -	Reducing in number rodents, development
Vormela peregusna pallidor	+ - - - -	Reclamation of virgin lands, application of che for extermination of rodents

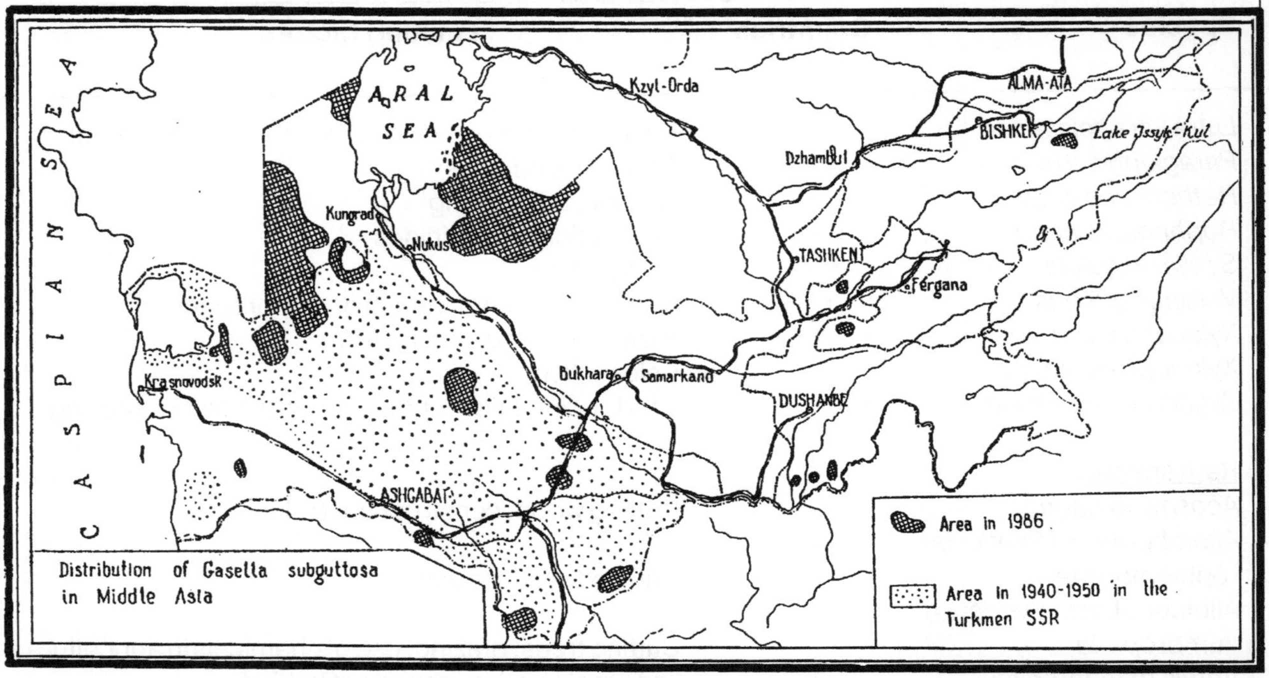

Figure 7.2. The area of dzheiran in Central Asia (Rustamov, 1986)

Protection of animals in the reserves

Many animals are protected in reserves of Central Asia (Neronov & Bobrov, 1990). The list of reserves is given in Table 7.2.

But these reserves have a miserable existence. For example, in Turkmenistan reserves occupy about 1.1 mln. ha. The local administration tried to break up these areas (Seitkurbanov, 1991). Officers of these reserves have a low salary, they didn't have scientific equipment and all their activity was based on the enthusiasm of activists of the green movement. They couldn't control the territory of reserves.

The reserve Tigrovaya Balka ("Tiger Ravine") occupied a unique area of subtropical jungle in the Vakhsh River valley of Tajikistan. Dense thickets of trees, shrubs and lianas were the habitat of many wild animals (pheasant, deer, hyena, wild cat and other rare animals). During the Soviet period many VIPs shot rare animals here. The size of this unique reserve was 70 thousand ha in 1970, then it was reduced to 48 thousand ha. In 1985 the government wanted to reduce the size of the reserve again (Illesh & Surkov, 1985). In the same year the fence surrounding the reserve was broken and local people began haymaking. Additionally irrigation water conducted from fields flooded the reserve (Illesh, 1985).

The number of wild ungulates in Central Asia has been reduced during the last century.

Dzheiran

Travelers of the past crossing Central Asia deserts always commented about the numerous flocks of dzheiran or gazelle *Gasella subguttosa*. Shooting dzheiran was very popular even in the beginning of the 20th century. The meat of dzheiran has a medical properties according to the belief of local people. It was sold on black market at high prices. Introduction of motor vehicles stimulated the near-extermination of this animal. A high fine was introduced for shooting dzheiran. It totaled in 1986, 800 roubles per animal (Flint & Prisyazhnyuk, 1986). But the enforcement of these regulations was so rare that many poachers were not punished. The same authors found that in the 1950s the northern limit of dzheiran range reached 48 degrees North, practically at the northern limit of the desert zone.

Table 7.2. Reserves of Central Asia (Neronov & Bobrov, 1990).

Reserves	Size, ha	Protected ecosystems					
		Mountain	Steppe	Semi-desert	Desert	River valley	Aquatiq
Kazakhstan							
Alma-Atinsky	73,320	+	+			+	
Barsakel'mess	18,300						+
Usturt	223,000				+	+	
Tajikistan							
Tigrovaya Balka	47,409		+	+	+	+	+
Turkmenistan							
Amudarya	50,000				+	+	+
Repetek	34,600				+		
Badkhyz	87,600	+			+	+	
Kaplankyrs	570,000				+		
Krasnovodsk	262,037						+
Synut-Khasardag	29,700	+				+	
Uzbekistan							
Aral-Paigam-Bar	3,094					+	+
Badai-Tugai	5,929					+	+
Kyzylkum	3,985				+	+	

In the seventies, this limit moved southward by 300 km.

Rustamov (1986) published very interesting information on the population of dzheiran in Turkmenistan. In the beginning of the last century, dzheiran lived in an area of 370 thousand km², 76.4% of the territory of the republic. Its population was 250 – 300 thousand head. In the forties, flocks of dzheiran were re-duced to 100 thousand head. In 1950 hunting dzheiran was forbidden, but its population decreased because of poaching. In 1986 not more than 10 thousand head of dzheiran remained in Turkmenistan (Figure 7.2).

Reduction of the dzheiran population has led to reduction of the number of gepard; this beast of prey disappeared from Turkmenistan. In such a way, the people destroyed the trophic chains the future effect of these changes could not be foreseen. Regeneration of the dzheiran's population can be achieved not only in situ but also *ex-situ* (Figure 7.3). Successful experiments on breeding this animal were conducted in Gyaurs nursery of Turkmenistan. Conservation of dzheiran in situ met difficulties because of severe environmental conditions. In 1986, 700 heads of dzheiran died because of shortage of water in Badkhyz reserve of Turkmenistan. They tried to drink saline water and perished.

Saigak

Kazakh wild sheep or saigak (*Ovis ammon collium*) is the subject of merciless shooting by poachers. In the past this animal was common in the drylands of Kalmykia, Kazakhstan and Mongolia. In the 1930s the number of saigak was severely reduced. Then strict measures were taken on conservation of saigak in the former USSR. The population was restored and the number of these animals on the USSR territory totaled 6 mln. head in 1982, then decreased to 1.7 mln. head in 1983 and to 1.0 mln. head in 1984 (Pral'nikov, 1985). Shooting saigak from cars is especially destructive. Saigak could run with a speed of 60 km per hour during 20 minutes. Then the animal dislocated its legs and

Figure 7.3. Young dzheirans. Photo by A.Guseinova

poachers gained the victory... The fate of saigak worsened in the newly independent countries of Central Asia. Poaching rose to an unprecedented scale. Poachers drove saigaks to special fenced sites and sawed off their horns! These horns have high value in oriental medicine. They were transported to other countries bordering Turkmenistan (Smirnov, 1993).

Kulan

Onager (*Equus hemionus onager*), also known as kulan, was in the past a typical beast in Central Asia. Brem (1902) said:"Kulan was found in southern Siberia, in Turkestan, in the desert steppe of Mongolia and in the Tibet mountains".

The population of kulan on the territory of the former USSR, was reduced and in 1941 only the one autochtonous population of kulan in the Badkhyz reserve totaled 250 head. In 1990, thanks to special measures on protection of this rare animal, its population reached 3,000 head (Sokolov, 1990). According to Neronov & Bobrov (1990), the population of kulan totaled 2,000 head in the Bagkhyz reserve. Above that, two new populations of kulan were created by artificial transmigration in the Barsakel'mes Island (200 head) and in the Kopetdag game sanctuary (50 head).

At present a new problem has arisen of regulation of kulan numbers in the Barkhyz reserve. The further increase in number of animals could lead to zoogenic desertification of the reserve's territory because of limited forage resources. Kulan also can damage bordering agricultural lands.

7.2. Other species of wildlife stand also in need of protection

Rodents

Rodents can play an important role in desert ecosystems, in particular in the soil formation process and in vegetative cover development. In addition, some species of rodents can migrate to agricultural fields and can damage cultivated crops. By digging the holes, rodents move great quantities of soil material from lower to upper horizons. Nurgel'dyev (1969) told about formation of sandy mounds around rodent holes in Karakum. Solonchak can arise in these places under the action of atmospheric precipitation if sand contains gypsum or other salts.

In the Karakum desert, a mass outburst of populations of rodents is observed each 7 – 8 years. During these years, rodents can compete with sheep by consuming 15% of the green biomass. Mass propagation of *Rhombomus opti-mus* and *Meriones eritrourus* was registered in the deserts of Turkmenistan in 1950s (Tashliev, 1988). The number of holes per hectare totaled 8,000.

According to the same author, mass propagation of *Meriones maridinus* and some other rodents was observed in 1955 in the Karabil' district of the Turkmen SSR, on an area of 200 thousand hectares. The number of rodents reached 20,000 per hectare. It was a real zoogenic desertification. But nearly all rodents practically disappeared the next year. Scientists couldn't explain this phenomenon. Desert rodents could also migrate to agricultural fields and damage c orn crops.

Snakes

The number of snakes has diminished everywhere in Central Asia. There are three reasons of that:

1. Destruction of habitat in connection with new land reclamation,
2. Increasing human pressure. Man and snakes often run into each other. The results of this meeting is usually tragic for snakes,
3. Snakes have become the subject of exploitation. People "milk" them, taking the venom. This venom is used in medicine.

Cobra (*Naja oxiana*) and gyurza (*Vipera libetiana*) have suffered an especially great reduction in number. Cobra has been listed in "The USSR Red Data Book" and in "The Red Data Book of IUNC". According to "The Red Data Book of the Turkmen SSR (1985) the nubmber of cobra in Central Asia (Uzbekistan, Turkmenistan and Tajikistan) totaled 300 – 500 thousand individuals. The number of snakes has been diminishing in Karakum. As Kuleshov (1992) said, some years ago 5 individuals of cobra could be met in the desert per 1 km^2. At present this meetings are very rare.

Special serpentaria have been set up in Central Asia to take the venom of poisonous snakes. According to Tashliev (1988), 576 cobra and 2,668 gyurza were caught in the Badkhyz serpentarium during 1967 – 1979. Gyurza could give 0.1 –0.4 g of venom (in dry weight) per one "milking". The medical industry of the former USSR needed 10 kg of venom per year (Gladkov & Rustamov, 1975).

Numerous poachers learned the dangerous profession of "snake charmer". Since 1991 snakes have been exterminated without any pity. For example, in 1993 a group of poachers from Russia was arrested in Turkmenistan. They tried to transport to Russia about 30 poisonous snakes (Kuleshov,1993).

Fishes

In Central Asia there are many water bodies – large and small rivers, lakes, storage lakes and irrigation canals. The Aral Sea before it started to dry produced 40 – 50 thousand metric tons of fish annually. Among fishes there were some valuable species, for example *Cipriana caprio, Stizostedion lucioptera, Barbus brachicefalus* etc. At present, the Aral Sea has totally lost its potential as a fish-producing water body.

Several experiments were conducted in Kirghizstan, Tajikistan and other countries of Central Asia on introduction of new fishes. Salmon was introduced in the lake Issyk-Kul'. In the Pamir lakes, several species of *Corregonus* sp. and salmon were introduced (Babaev & Kharin, 1994b).

But the most interesting experiments on introduction of new species of fish were conducted in Turkmenistan (Kharin, 1995). After the construction of the Karakum Canal, its channel began to fill with hydrophylous and water plants (*Typha australis, Phragmites sp., Potamogeton lucens, Batrachium pachycaulis, Ceratophyllum demersum,* etc.). Vegetation became an obstacle for the movement of water. Furthermore, the quality of water became bad because of plant matter decomposition. So scientists of Turkmenistan introduced fishes from China into the canal, mainly *Ctenopharyngodon idella* (Russian name "belyi amur"). According to this project, new fish would destroy the canal's vegetation and the people would have benefit. The project was successfully realized. Together with belyi amur the spawn of a frog from Japan (*Rana nigromana*) was introduced to the Karakum Canal. Now this animal is a common species in the canal zone (Murkina, 1976).

Soon after the realization of the project, the negative consequences of fish breeding were discovered. Kogan (1991) found that the destruction of vegetation in the canal caused the intensive development of phytoplancton algae. These algae caused the biological contamination of water reservoirs. Shortage of oxygen in water near the bottom together with the increase of content of biogenic substances also caused the deterioration of water quality. The vegetation destroyed by new fish had in fact played the role of the biogenic filter and stimulated the purification of water. He gave the following examples: mild vegetation growth had a positive effect on the quality of water in the Kurtly reservoir, close to Ashkhabad, while the quality of water became worse after the breeding of belyi amur. Similarly the quality of water in the Gindukush reservoir, where fishes didn't breed, was good.

So it is clear that man plays an important role in the transformation of natural ecosystems. All measures effecting the change of flora and fauna must be kept under constant control. Only in this case can the equilibrium be maintained.

7.3. Hunting and poaching

In all civilized countries where the level of ecological thinking of population is very high, these two occupations stay on opposite sides of the law. In such countries, hunting is a sport or a legal occupation of people and poaching is a criminal deed. But the countries of Central Asia are not among those countries. Desert lands became the arena for activity of many poachers, from plain people to high-ranking functionaries. Poachers on motor cycles and in off-road cars penetrate to remote areas in the desert. Firearms can be purchased by all people on the black market, so poachers use rifles arm and even machine carbines.

De jure game-shooting on the territory of the USSR (including Central Asian republics) was regulated by several decrees, in particular "Seasons of game-shooting and license to firearms" (1919), "The law on protection and utilization of wild animals" (1920) and by several regulations issued by local administration. The constitution of the USSR, articles 18 and 67, also protected the wildlife.

But *de facto* these regulations were not always enforced. Sometimes high officials participated in a "king's chase". For example, the former Minister of Defense Marshal A. Grechko liked hunting in state reserves. In Kurgal'dzhinsky Reserve of the Kazakh SSR he shot white swans from helicopters. He was a marksman! (Vladimirov, 1993b).

During one month in 1993 hunting inspectors of Turkmenistan arrested more than 1000 poachers. Hundreds thousands of manats were paid as fines during the same year in the country (Musienko, 1994). Two young men were arrested in Kyzylarvat district because they shot young partridges. A big fine was paid by a certain P. Palvanov who shot two arkhar. A team of poachers shot dzheiran and hares in the Tashauz velayat.

Hunting inspectors often used physical strength to arrest poachers. Inspector G. Maksimenko beat severely a war invalid A. Tyuzhin who fished in a forbidden place, in the Kurtly storage lake, close to Ashkhabad. The poacher had an artificial leg and couldn't resist. The Association of War Veterans of the Turkmen SSR tried to put this obstinate inspector on trial, but he was not punished for exceeding his authority (Polovinkina, 1988a).

We don't have complete figures either on poaching or on the real number of animals shot "by legal grounds". Table 7.3 gives data on furskins purchased in the Turkmen SSR by state agencies during 1924 – 1985. These data are not complete because part of the skins are usually dressed by local fur-dressers. Above that, a considerable part of fur-skins was sold in the black market. That occurred because of low prices for skins in state agencies. We don't have any statistics on this subject. Is that plenty or scarce animals or it is only a small part of numerous populations? We can't answer this question. But the absolute figures are great. Really, the mean annual number of purchased skins in Turkmenistan totals 57,694. Probably for the five countries of Central Asia the number of purchased skins will be five times greater.

Table 7.3 show the annual dynamics of purchasing. The author of this publication gave only some figures on the annual numbers of purchased skins. For example, during 1970 – 1985 the following were shot annually: 341 wolves, 3,526 jackals, 12,576 foxes, 332 corsacs and 957 wild cats.

Table 7.3. Number of fur-skins purchased in Turkmen SSR during 1924 – 1985 (Tashliev, 1988).

Category	Animal species	Number
Main species	Wolf	26,070
	Jackal	170,160
	Fox	1,298,800
	Korsac	00,430
	Wild cats	247, 800
	Hare	426,618
	Coypurat	16,846
	Ondatra	6,746
	Polecat	4,267
Subtotal		2,297,737
Secondary species	Musteline	10,692
	Weasel	436
	Bodger	3,665
	Gofers	983,774
	Jerboa	54,883
	Other rodents	163,603
Subtotal		1,217,053
Rare species	Tiger	3
	Gepard	3
	Leopard	420
	Lynx	153
	Karacal	467
	Manul	100
	Hyena	532
	Marten	2,700
	Otter	110
	Bear	8
	Honey badger	55
Subtotal		4,557
Total		**3,519,347**

We don't know how many animals could be shot without damaging their regeneration. Nobody could answer this question. And what is more, different scientists give different recommendations. For example, Gladkov & Rustamov (1975) found that jackals should be exterminated because they harmed breeding fur animals and poultry farming. Tashliev (1988) didn't demand the extermination of jackals. According to his opinion, the population of jackal could be regulated, and 8 to 9 thousand of this beast could be shot annually in Turkmenistan.

Marmot numbers were severely reduced in Kazakhstan. The marmot was known as "bai-bak", that means "sleepyhead". Prokhorov (1983), a correspondent of "Pravda", told a very interesting story about shooting marmot in Ka zakhstan. This animal had valuable fur and its fat was used in folk medicine. A gang of poachers headed by a certain "Pasha Millionaire" shot thousands of marmots. Poachers used rifled arms with telescopic sights, they shot marmots from the car, etc. They were not afraid of punishment because they had high patrons – F. Grishai, Chief Hunting Inspector of the Kazakh SSR and A. Zhumadilov, Head of the Department of Reserves of the Kazakh SSR. But at last, local hunting inspectors instituted proceedings against poachers. They demanded compensation in the sum of 475,000 roubles. But corrupted courts confirmed only 25,000 roubles. The sentence was mild; some poachers were acquited. After this scandal, Grishai was demoted. He got a new duty-Director of the Balkhash State Reaserve. He published an article in a local newspaper entitled "Struggle against poachers"!..

Activists of the green movement tried to cteat public condemnation of poachers. For example, Kurbanov (1979) proposed to discuss each case of poaching at special meetings of "labor collectives" in organizations in which the poachers were on the staff. But we didn't know a single case where a poacher had been "re-educated".

The killing of wild beasts was aggravated after the proclamation of the independence by Central Asian republics. T. Yagshimuradov, Director of the Badkhyz State Reserve in Turkmenistan, said: "In connection with market relations, poaching has been increased enormously and damage from it is great. And the fine for shooting wild animals has been increased. In1993 it totaled: 540,000 roubles for kulan, 390,000 roubles for dzheiran, 660,000 roubles for arkhar (mountain goat). In spite of that, poaching continues" (Kuleshov, 1993).

Turkmen police arrested on 18 January 1995 a group of poachers in Balkhan velayat of Turkmenista. Police discovered in two lorries the carcasses of 248 saigaks hot in Turever area, close to the territory of Kazakhstan. Five poachers were responsible for that. Then the third lorry was also stopped with 30 carcasses of saigak. Ch. Abaev, the head of police of this velayat, said that it was the most largest case of poaching in this region. Police instituted proceedings against these criminals (Ustaev, 1995b).

Many rare animals and plants of the Badkhyz Reserve remain on the verge of disappearance. The territory of the reserve is suffering from the impact of local populations who try to graze domestic animals, collect pistachio nuts, etc. Human activity becomes more and more active in the marginal zone, new land being developed for agriculture. So, many species of flora and fauna need special measures for their conservation. Kuleshov (1993) described in the following way the situation of protection of wild animals in this reserve. Poachers shot all kinds of animals: dzheirans, kulans, arkhars, and caught snakes for venom. A group of poachers from the Russian city of Perm tried to transport to Russia many poisonous snakes. Rangers confiscated 30 snakes and gave them back to the serpentarium of the Badkhyz reserve.

Yakshimuradov, director of the reserve, described the situation in the reserve which was founded in December 1941. The territory of the reserve is a unique place, with many animals and plants from "The Red-Data Book". Wild animals identify themselves the places where they are protected. For example, groups of kulan escaping from poachers try to run back to open-air cages constructed in the reserve. In these cages they feel safe.

Experts believe that the transition to market economy gave an additional impetus to further increase poaching. Great fines were introduced to improve the situation: for shooting kulan, 540,000 roubles; for shooting dzheiran, 390,000 roubles; for shooting arkhar, 660,000 roubles (at the rate of 1993). But some scientists believe that probably in the future people will see some of these beasts only in zoos.

7.4. Conclusion

The Red Data book of the USSR includes 5 animals which have probably disappeared, 21 endangered animals and 22 rare animals.

Destruction of natural habitats was especially harmful for conservation of wildlife. Tugai, this jungle of Central Asia, was always a refuge for many wild animals, birds and insects – the pollinators of higher plants. At present days tugai forests occupy some 40,000 hectares. Their area was considerably reduced during the last 20 years.

Reclamation of new lands and increasing human pressure on the environment have a negative effect on wildlife. Wild animals lose their habitat and perish by contacts with man.

Construction of the Karakum Canal and many irrigation canals transformed the desert environment. Special measures should be taken for conservation of flora and fauna in the canal bordering areas.

In 1991, or even two or three years before, "the troubled times" began, old regulations and old traditions were forgotten or lost their validity. Wild animals "were deprived of all their rights", speaking in allegoric language. But the worst is that these "troubled times" are going on.

New regulations should be developed for conservation of wildlife in Central Asian countries. But these regulations can give a positive effect only if plain people change their minds and nature conservation ideas spread over various strata of society.

Some people became poachers not only because of their irresponsibility but also because of shortage of food. So, if the economic situation in Central Asia will be improved, that gives an additional chance for wild animals to survive.

CHAPTER 8

ACCUMULATED ECOLOGICAL PROBLEMS IN NEW INDEPENDENT COUNTRIES OF CENTRAL ASIA AFTER 1991

Andre Zeiten, andre Vögel!
Andre Vögel, andre Lieder!
Sie gefielen mir vielleicht,
Wenn ich andre Ohren hätte!

H. Heine

8.1. Sovereignty, poverty and refugees

Desertification as one of the Accumulated Ecological Problems (AEP)

Desertification is only a part of many ecological problems. Glantz (1995) introduced the term "Accumulated Ecological Problems" (AEP). He included in the list of AEP such problems as acid rain, desertification, drought, deforestation, climate change, etc. Each of these problems is a result of a long accumulative process which can't be identified by the people at once.

Scientific incertitude and the lack of clear criteria are often obstacles for decision makers to prevent further worsening of the ecological situation. So, politicians can always find a pretext for their inertness in nature conservation activities. They can use only selective information published by some scientists to make a wrong decision. As to Central Asia, a malicious will of political leaders also played an important role in land reclamation and nature conservation.

Desertification, being a destructive process, hits not only the physical but also the social environment. From the social point of view, desertification begins when consumption exceeds production. The gap between production and consumption increases more and more as desertification progresses. This gap can be filled by three ways: decrease of consumption, external aids and increase of production. Decrease of consumption can be achieved first of all by regulation of population growth. Increase of production in its turn can be achieved by melioration of degraded land. External aid can only help the local people to survive, but this approach is hopeless from the strategic point of view.

New sovereign states of Central Asia

The USSR, the mighty state of workers and peasants, "The Empire of Evil", disintegrated in 1991. New states appeared on the ruins of the empire. Kazakhstan, Kirghizstan, Uzbekistan, Turkmenistan and Tajikistan proclaimed their independence. But they did not win their sovereignty in a struggle, it was granted by a TV program!..

The communist party of the USSR was disbanded. Instead, new parties named "Democratic" were set up in the Central Asian countries. Communist ideology was replaced by Moslem ideology. Many new mosques were built; for example a grandiose mosque was built in Geok Tepe, where Russian troops massacred thousands of Turkmen in 1881. More than 1,000 mosques were built in Kirghizstan during the last 5 years. In spite of that, the Christian church is becoming more and more popular among young Kirghizes. E. Kaptagaev, the head of the Department of Cults of this country, said that 24 religious communities of Protestants were registered in Kirghizstan. He noted: "The Kirghizes being Moslems, considered this people as heretics. They stay under the threat of physical violence or could be banished from villages" (Anonymous, 1996).

Hypertrophied feeling of the national self-consciousness and sovereignty became mighty factors in the political development. The ardor of sovereignty in some cases was transformed grotesquely. In the old Communist tradition, new independent countries paid much attention to so called "visual propaganda". For example, a correspondent of "Komsomol'skaya Pravda" has seen in Tashkent a strange carved image - an enormous terrestrial globe with the map of Uzbekistan. He asked a policeman on duty: "What does it mean?", and the policeman said: "Why, don't you understand? It is the terrestrial globe of Uzbekistan" (Grineva, 1992).

Kazakhstan has a claim on a leading role in the region. The leaders of the country want to transform Kazakhstan to a "Central Asian Snow Leopard" with a strong political leader (Ardaev, 1997). President N. Nazarbaev has developed a program of transition to "a new healthy way of life". As he said, an additional tax would be paid by married couples who don't want to have children... In this way he paid tribute to Moslem ideology. Addressing the poor and unemployed he said : "You are not forgotten!.. I sympathize with you in your sorrow and understand your severe situation. I call you: Instill into you confidence in the future". But according to his words, the financial situation would not be improved during the next four years. And the President asked poor people "to stand more during some time", covering their needs, without any help of the state.

Uzbekistan is the second country having a claim on the leading role in the region. Azizova & Khasanov (1999) cited the words of M. Karimov, President of Uzbekistan: "Strong authority means a strong state". But the authors of this publication said" The political regime in Uzbekistan is authoritative", they also called the economic system of the country as "state capitalism of the oriental type". Probably, "the oriental features" include first of all the social structure of the society. Uzbekistan is the country of young people. Of the 24.5 mln. population (1999), about half is young people, 18 – 19 years old. So, only 1/3 of the population are employed workers who pay taxes. Azizova & Khazanov (1999) contened, that the economy of the nation was developing successful by the guidance of the government. They said, "the population has a low but stable standard of living" (?).

As known, the planned economy of the former USSR was far from being perfect. But disintegration of the Soviet empire caused the collapse of this system. Economic ties between the former Soviet republics were broken. The living standard of people in Central Asian countries is now 3 – 5 time lower than in Russia. The minimum monthly salary in the region was in 1994 (USD): 0.99 in Kazakhstan, 3.75 in Kirghizstan, 11.70 in Uzbekistan, 15.0 in Turkmenistan, 1.37 in Tajikistan. Minimum pension was also very low (USD): 2.49 in Kazakhstan, 5.0 in Kirghizstan, 12.7 in Uzbekistan, 5.0 in Turkmenistan, 4.37 in Tajikistan (Babaev & Kharin, 1994a). After the falling of the rouble rate in Russia on 17 August 1998, the coefficient mentioned above became 3 – 5 lower. That means that the living standard in Central Asia countries is very low compared with Russia. It is one of the reasons why millions of people from all CIS countries fly to Russia. According to published data (Anonymous, 2000), the mean nominal monthly wages in Kirghizstan totaled in October 1999 USD 23.9. But the real value of these wages fell during the same month by 4.4% because of inflation.

Corruption and shady economy

An outburst of criminal activities was a characteristic feature of all new countries formed on the territory of the former USSR (Nadein, 1985). The shady economy developed during perestroika, stimulated the concentration of capital in a small number of people. The second group of rich people was formed by corrupted bureaucrats. In many cities of Central Asia, quarrels of criminals, accompanied by shooting, assassination of businessmen and officials resembled the life in Chicago during the thirties.

Bribes and baksheesh became a standard practice of new administration. For example, M. Akaev, president of Kirghizstan said: "Armed bands have been formed. We must pass a special decree against such criminal organizations. And what about the Ministry of Internal Affairs? Maybe the Ministry itself is forming these bands". Police of this country suppressed 100 bands of criminals during 5 months of 1996. But the worst was, that the people didn't blame gangsters. Some people considered them as tough dodgy guys who could "settle their affairs" under severe social conditions that prevailed (Ryabushkin, 1993).

A. Akaev told about the activities of criminals in his country (Anonymous, 1999a). Such economic crimes as non-payment of taxes, bribes, foundation of fictitious companies, legalization of money earned by illegal activities, etc. are common. In his opinion, the so called "shady economy" totaled 10 – 20% of the gross national output. According to other sources these figures totaled 20 – 25%. The latter were very close to that given by Nishanov (1988b) who described the bribes and pripiska in Uzbekistan during the Soviet period. Probably the change of political system in the former Soviet republics didn't improve the honesty of the powers that be!..

Uzbekistan lost commodity markets in Russia because of high custom tariffs and bribes demanded by road police during transportation of agricultural products. So, many farmers curtailed agricultural production and preferred to grow poppy and hemp. Unemployment in the Fergana valley totaled 80% (Vinogradov, 1991). As Le Vain (1 997) found, bribes, corruption and bad management became real obstacles for economic development. Foreign investors reduced their activity. Many foreign businessmen lost their money and left Uzbekistan. Agricultural production in the country was insufficient. Low yields of cotton and low prices on row cotton on the international market caused damage totaling 400 mln. dollar during 1996. On top of that, a decrease in production of wheat forced the government to purchase grain abroad for 200 mln. USD.

The production and delivery of narcotic drugs to Russia and other countries became a legal business in Tajikistan. The head of he Tajik mafia, nicknamed "Lesha Gorbun" controlled this "branch" of the economy. He had an armed guard and laboratories for processing opium poppy and producing heroin (Mlechkin, 1993). Thousands of women and children worked on poppy plantations. Poverty, unemployment and the wish to survive forced them to participate in this criminal activity (Karpov, 1990).

Many thousands people were killed in the civil war in Tajikistan. In 1997 civil war in this country was stopped and the first signs of economic stabilization were registered (Kidenis, 1997).

According to Kuleshov (1993), poppy is also cultivated in the Chu valley in Kirghizstan and in Karakalpakia. The so called the "Green Triangle" is functioning in the Serakhs district of Turkmenistan, at the junction of the borders of three countries: Turkmenistan, Afghanistan and Iran. Smugglers deliver drugs from Afghanistan to Iran through the territory of Turkmenistan by a secret path. The governments of Turkmenistan and Iran signed a special agreement for eradication of this criminal activity.

Violation of human rights in the new independent countries was dicussed many times in the mass media of Russia. For example, ethnic Russians, Kossaks from Semirechie region, tried to set up a public organization which could protect their rights. But Kazakh police arrested their head ataman N. Gun'kin. N. Sidotova, the head of the Russian Center in Alma Ata was charged with resistance to police and the law court passed a verdict of guilty on her (Ardaev, 1997). The situation in other countries of Central Asia is not better.

According to a special request conducted by the European Bank of Reconstruction and Development, the level of corruption in the former Soviet republics was very high (Down, 1997). Businessmen working in new independent countries met corruption everywhere, as the request of 3,600 respondents confirmed. For example, they were asked: "How often were you compelled to make an additional payment ?" For 10 countries of CIS, the answer of 65% of respondents was: "Often, very often or always". As a result, the complex criterion of corruption was very high, 15% higher compared with African countries located to the south of Sahara, and on 24% higher than in Latin American countries.

As this report said, corruption in the CIS became a stable and well organized business. Any businessman knew how much he had to pay, but the state policy in these countries is not stable and all regulations and the rules are unpredictable. For sustainable development, the institutions in these countries responsible for support of the market economy must remove all bureaucratic barriers and intensify the struggle against corruption.

Refugees and displaced persons

Millions of Russians and other non-Moslem people left the territory of Central Asia after 1991. The main causes of this migration are the following: violation of human rights, poverty, racial discrimination, bad outlook for young generation and worsening of ecological conditions. The real figures of the people who became refugees are unknown. By rough estimation, they totaled 4 to 5 mln. Neither the Russian government nor the governments of Central Asian countries wanted to emphasize the problem of refugees. As a German proverb said, "Sie machen gute Miene by schlechter Spiel". That means "They put a good face on the matter". It should be noted here that many people of local nationalities also migrate to Russia looking for jobs and better social security.

Mass migration of Germans was registered in Kazakhstan. From 700,000 Germans living in this country, 300,000 left Kazakhstan in 1993. The rest wanted to do that next year. According to Glazovsky & Shestakov (1995) the annual rate of migration in Kazakhstan totaled 1.02 – 1.18 million people per year. Among them 600 – 700 thousand leave the country, the rest move within the country. Baitullin & Bekturova (1997) give the following figures on migration in Kazakhstan. 300,000 people in Kazakhstan, including 20,000 from the Aral Sea region left their residences in 1995. The flow of refugees from other countries of Central Asia also has been continuing.

The trend is observed of migration of the local people from rural country to towns where living conditions and social service are better. Table 8.1 gives information on movement of people in zone of ecological disaster.

The situation in Tajikistan became especially unbearable. Of 300,000 Russians registered in the country, 250,000 left by 1996. The rest, mainly poor people and invalids, as Malikov said (1996): "...are slowly dying from hunger and cold in their unheated dwellings". Priests of the Russian Orthodox Chirch performed a funeral service for the deceased and the soldiers of the 21st Russian division provided them free with coffins and transportation vehicles to the cemetery.

V. Nekrasov, a geologist who had worked 34 years in Tajikistan told about many difficulties met by refugees. He left Tajikistan for Russia. He said:

Table 8.1. Increase on town population in zone of ecological disaster, in thousand (Babaev & Kharin, 1994b).

Towns	1959	1970	1980	1989
Kzyl-Orda	66	122	157	153
Nukus	39	74	109	169
Urgench	44	76	100	128
Tashauz	38	63	84	112

" For transportation of my belongings and crossing the state borders I paid officially 1 mln. roubles, above that I paid 2 mln. roubles baksheesh" (Malikov, 1996).

Who are the refugees? United Nations bodies, in particular, the United Nations High Commissioner for Refugees (UNHCR) defines as "refugees those people who are outside the country of origin because of the fear of persecution for reasons of race, religion, nationality or political opinion". As Watanaba (1995) said, in the course of time the scope of this definition has been broadened. People fleeing from violence, internal conflicts or massive violation of human rights could be also considered as refugees. However UNHCR doesn't use the term "refugees" for persons who left their countries under the pressure of economic or environmental constraints. In this case the term "displaced persons" is used.

Cardy (1994) gives the definition of environmental refugees adopted by UNEP : "Environmental refugees are those people who have been forced to leave traditional habitat temporarily or permanently because of marked environmental disruption (national and/or triggered by people) that jeopardized their existence and/or seriously affected the quality of their life". In 1988 about 10 mln. people could be considered as environmental refugees. They could be temporarily or permanently displaced because of environmental stress. Some of them migrate beyond the native country borders without any hope of returning home.

The most detailed classification of people who move was given by Stiles (1995). This classification is given in Table 8.2.

Table 8.2. Categories of people who move (Stiles, 1995).

Categories	Natural force	Human coercion	Voluntary
Internal	Displaced person	Repressee, oustee	Migrant
International	Displaced person	Refugee (asylum seeker)	Migrant (asylum seeker)

"Asylum seekers" are a temporary category. They are usually registered as "refugees" or "migrants" in receiving countries. The author of this classification added two other descriptors to these categories: "documented" or "not documented" and "temporary" or "permanent". The same author give the following information about the people who move at a global scale (Table 8.3.).

The author didn't say if the refugees of Central Asia were included in this table.

People leaving Central Asian countries don't know this terminology. The main part of them belong to categories of "not documented" and "permanent". They only have *the status of human*. In some countries of Central Asia Russian embassies give small grants-in-aid to those Russians who want to migrate to the motherland. They belong to the category "documented". For example, the Russian embassy in Ashkhabad has a long waiting list of the people who want to migrate to Russia. But the people have to wait 2.5 – 3 years because the embassy accept only a limited number of people every day.

Table 8.3. Number of peoples who have left their place of origin (Stiles, 1995)

Categories	Number, mln.
International migrants	100
Refugees	25
Internally displaced	27
Displaced by development	70
Internal migrants	1,000 ?

The office of UNHCR was opened in Moscow in 1992. Sadako Ogata, the head of UNHCR, visited Moscow in 1996, and was interviewed by a correspondent of "Izvestiya" (Mikheev, 1996). She said: "The world's community has needed four years to be aware of the problem of refugees in Russia and in the former republics of the USSR". It seemed that UN agencies didn't observe the movement of millions of people to the frontiers of Russia. The greatest exodus of the 20th century was overlooked for years... The UNHCR office in Moscow funded 35.8 mln. US dollar as aid to refugees in Russia during 1992 –1996. It was absolutely clear that this small amount of money couldn't solve the problem of refugees in Russia. Above that, Ogata said, these funds were given *to the government*, but *not to refugees directly* (Italics is our). So, corrupted bureaucrats could misappropriate "the lion's share" of these funds.

Many years ago Lenin said, that "Russia was a prison for the people". After 1991 Russia transformed to asylum for all people of the former USSR (not only Russians) seeking refuge here. Many cities of the southern part of Russia and even Moscow are overcrowded with refugees from Central Asia and the Caucasus. They include all categories of people who move. It is striking that these people belongs to many nationalities and ethnic groups not common for Russia itself. That is a real Babel!...

According to the forecast of the Russian Migration Program, an extra three million immigrants will come to Russia by the beginning of the 3rd millennium. It is expected that these refugees will come from Central Asia and other countries of CIS and a comparatively small number of people will come from so-called "distant foreign countries" (Churakova, 1998).

8.2. Scarcity of water resources

Water resources of Central Asia

Water always played an important role in the life of people of Central Asia. Irrigation agriculture appeared here 4,000 years ago. It was always the main occupation of the people in oases. People could construct dams and conduct water by irrigation canal (aryk) to the fields. Underground tunnels (karez) intercepted underground water at the foots of mountains. Desert shepherds drilled deep wells by primitive tools. But the scarcity of water resources always limited not only development of agriculture but also stock raising.

At the present time, the increased population of Central Asia needs more animal products. But the increase of livestock numbers can be achieved only by better watering of desert rangelands. Additional forage can be also obtained by cultivation of forage crops on irrigated farmlands.

During the Soviet period, industrial development became the third domain of human activities in the region. New industrial centers need more and more water for technical purposes and for human consumption.

Misuse of water resources is one of the main causes of land degradation in the region (Kharin, 1996a). Water scarcity in Central Asia is the key constraint to sustainable development. The problem is aggravated by the fact that shared water resources need joint efforts of five countries for their management.

A considerable part of Central Asia belongs to the Aral Sea basin. Two big rivers of Central Asia, the Amudarya and the Syrdarya are exotic rivers. They cross the plains of Central Asia but their sources are located in the mountains, rich in precipitation and glaciers. Water resources of Central Asia were studied by many scientists. They could be divided into two parts:

1. Active waters which could be used directly. River flow is a considerable part of these resources.
2. Potential water resources which include ephemeral run-off from takyr surfaces in the desert and temporary flow of dry water channels. Their sources are located in mountains rich in precipitation and glaciers.

According to the rough estimation of Shul'ts (1985), river flow in Central Asia totaled 155 km^3 per year. Voropaev (1982) gave flow figures in the amount of 127 km^3 per year. Kirsta (1991) said that the plain part of Central Asia had water resources of 127 – 129 km^3 per year. But in our opinion the most accurate information on water resources in Central Asia was given in the paper published by Raskin, Khansen, Zhu & Stavitsky (1993). Table 8.4 gives information on surface waters of these two big rivers.

Sources of underground waters are indicated in Table 8.5. They total in the Aral Sea basin 12,300 mln. m^3.

It is very difficult to estimate the real figures of potential water resources. By rough assessment of Leshchinsky (1974) the resources of ephemeral runoff in Central Asia totaled 431 mln. m^3 per year. But Kirsta (1991) said that many takyr sites were used during the last years for cultivation of agricultural crops or they were covered with desert vegetation. So they could not be used as water harvesting sites, and the potential water resources should be reduced to 350 mln. m^3 per year. Above that, on piedmont plains of Central Asia surface flow could be formed in dry river channels by rain water runoff. Only in Turkmenistan according to Kirsta (1991) it totals 50 mln. m^3 per year.

Projected water demand

According to estimations of experts (Raskin, Khansen, Zhu & Stavitsky, 1993), the projected water demand in the Aral Sea basin totals 97.3 km^3 per year from which 82% goes to agriculture, 8% to industry, 6% to municipal use, 3% to fishery and 1% to stock raising. Cotton is the main agricultural crop cultivated in the region. The figure of 97.3 km^3 relates to the final use of water on farms, industrial enterprises, etc. But the real consumption of water includes the loss of water for evaporation, infiltration etc. So these figures should be increased by 30%, to 127 km^3 (Table 8.6).

Table 8.4. Sources of ground waters in the Aral Sea basin (Raskin, Khansen, Zhu & Stavitsky, 1993).

Irrigation systems	Annual flow, mln. m^3	Irrigation systems	Annual flow, mln. m^3
The Syrdarya basin		The Amudarya basin	
Toktogul	12,797	Vakhsh	21,665
Karasu, left	0,470	Gyandzh	37,602
Karasu, right	1,809	Kunduz	3,535
Shaidansai	0,112	Kafirnigan	7,741
Karadarya	11,284	Surkhan & Sharabad	4,757
Kassansai, reservoir	0,301	Murgab	1,264
Kurvasai	0,284	Tedzhen	0,191
Shakhimerdan	0,869	Atrek	0,212
Isfara	0,335	Kashkadarya	1,431
Sokh	0,490	Guzardarya	0,203
Aksu	1,512	Zeravshan	5,716
Chakyr	12,911		
Arys' and Bugun	2,658		
Low Syrdarya	0,464		

Table 8.5. Sources of underground waters in the Aral Sea basin (Raskin, Khansen, Zhu & Stavitsky, 1993).

Syrdarya basin	Mln. m^3	Amudarya basin	Mln. m^3
Vysoky Naryn	1,000	Pyandzh	0,173
Fergana Valley	4,800	Vaksh	0,275
Syrdarya, middle flow	1,000	Kafirnigan	0,459
Chakyr	1,000	Kashkadarya and Karshi	0,299
Artur	0,250	Zaravshan – Bukhara	1,030
Syrdarya, low flow	0,250	Chardzhou	0,414
		Karakum	0,591
		Amudarya, low flow	0,341
Total	**8,300**	**Total**	**4,000**

The same scientists have developed a model of water flow to the Aral Sea from two big rivers of Central Asia. The model is based on studying the climatic trends for the next 25 – 30 years (Table 8.7). The size of the Aral Sea will be reduced from 40,780 km² in 1987 to 9, 411 km² in 2015.

The problem of water scarcity in the region will be aggravated in the beginning of the 21st century. During the Soviet period, several concepts of saving the Aral Sea were published, in particular Sultangazin (1991) proposed three possible approaches in rehabilitation of the Aral Sea. They included:

1. Partial transfer of the flow of Siberian rivers to Central Asia (25 – 60 km³ per year),
2. Cloud seeding in the basin of the Aral Sea,
3. Transfer of the Caspian Sea water to the Aral Sea (about 30 km³ per year). A new geopolitical situation was formed in the region after 1991 and all such concepts fell to oblivion.

The new independent countries started efforts to solve the problem of the Aral Sea. Several national, non-governmental and inter-governmental organizations held several meetings and symposia, but in vain... The solution of the problem was not found. Yosikiro Takano (1993) identified three tasks which should be solved by rehabilitation of the ecological situation around the Aral Sea:

1. The main task, rehabilitation of normal ecological situation, including the reconstruction of irrigation network.
2. The urgent task, provision of people with potable water and improvement of medical service.
3. Supporting measures, reorganization of national and local self-government institutions, training, research, etc.

The State Committee on Nature Conservation of Uzbekistan elaborated a concept of economic development in the Aral Sea region (Khabibulaev, 1993). According to this concept, the problem could be solved only on the basis of a multilateral agreement of all countries of the region and Russia. Full rehabilitation of the Aral Sea to its size in

Table 8.6. Projected water demand in the Aral Sea region, km^3 (Raskin, Khansen, Zhu & Stavitsky,

Irrigation systems	Agriculture	Industry	Municipal	Stock raising	Fishery	Total
			The Amudarya basin			
Pyandzh	1.11	0.05	0.22	0.02	0.02	1.42
Vakhsh	2.48	0.08	0.30	0.03	0.19	3.03
Kafirnigan	1.40	0.03	0.25	0.01	0.45	2.14
Sur-Sherabad	3.19	0.00	0.27	0.01	0.03	3.50
Afganistan	0.00	0.00	0.00	0.00	0.00	0.00
Karakum	7.22	1.69	0.41	0.02	0.07	9.41
Kashkadarya	5.56	0.05	0.40	0.01	0.05	6.07
Bukhara-Zeravshan	8.89	0.66	0.65	0.11	0.04	10.35
Chardzhou	2.39	0.17	0.16	0.00	0.09	2.81
Khorezm	2.60	0.06	0.15	0.00	0.05	2.85
Tashauz	2.59	0.00	0.05	0.00	0.05	2.69
Kar	7.37	0.20	1.12	0.03	1.51	7.23
Total	44.80	2.98	2.98	0.24	2.55	53.55
%	84	6	6	0	4	100
			The Syrdarya basin			
Vysoky Naryn	2.11	0.01	0.03	0.00	0.07	2.22
Fergana valley	12.48	0.31	1.31	0.00	0.07	14.17
Syrdarya, middle flow	7.45	2.14	0.33	0.00	0.12	10.04
Chakyr	5.17	2.30	1.39	0.00	0.12	8.98
Artur	2.11	0.21	0.15	0.00	0.12	2.59
Syrdarya, low flow	5.26	0.07	0.11	0.00	0.33	5.77
Total	34.58	5.04	3.32	0.00	0.83	43.77
%	78	12	8	0	2	100
Total in the Aral Sea basin	**79.88**	**8.02**	**6.30**	**0.24**	**3.38**	**97.30**
%	**82**	**8**	**6**	**0**	**8**	**100**

Table 8.7. Annual flow of the Amudarya and the Syrdarya to the Aral Sea, km^3 (Raskin, Khansen, Zhu & Stavitsky, 1993).

Years	Amudarya	Syrdarya	Total
1987	6.28	2.52	8.79
1990	1.88	2.13	4.01
1995	0.69	1.23	1.92
2000	1.84	1.08	2.92
2005	1.09	0.97	2.06
2010	1.79	0.83	2.62
2015	1.03	0.57	1.60
2020	1.75	0.16	1.91

1965 is practically impossible. The author said that sustainable development of the region is impossible without additional water resources transfered from outside. But it was not clear how the problem of scarcity of water resources would be solved? Under the present geopolitical situation, transfer of water resources from Russia is possible but hardly probable.

8.3. Worsening ecological and social conditions in the areas affected by desertification

Fall of living standard

The living standard of the people in new independent countries of Central Asia is characterized in Table 8.8 below. One can see that, for example, consumption of grain products in 1991 was lower than in the beginning of 20th century. Selective questioning of populations conducted in 1990 show that provision of people with main food products was insufficient from their point of view Table 8.9).

The situation with production of grain was studied by Ministry of Agriculture of USA (Bablumyan, 1999). According to this information, the share of Russia and Kazakhstan in production of grain was falling. On the forecast the gross output of grain in Russia was expected to be 57.2 mln. MT (mean yield 1,140 kg/ha), the same for Kazakhstan was 8.2 mln. MT (mean yield 740 kg/ha), in Ukraine 27.4 mln. MT (mean yield 1,320 kg/ha). The share of Russia in grain production in 1999 totaled 52 in CIS (55.3% in 1998), the same for Kazakhstan was 7.4% and 7.9%, the same for Ukraine totaled 24.5% and 23.4%. According to estimates of experts, production of grain in Turkmenistan and in Uzbekistan increased during the last years.

Table 8.8. Consumption of food per inhabitant, kg/year in Central Asian countries (Economy of Commonwealth Countries, 1993).

Countries	Year	Meat	Fish	Milk	Egs, peaces	Sugar	Vegetable oil	Grain products	Potatos	Vegetabls	Fruits berries
Kazakhstan	1985	48.0	10.9	260.0	217.0	37.1	9.8	146.0	89.0	90.0	22.0
	1992	63.0	7.0	270.0	180.0	26.2	9.0	160.0	80.0	64.0	17.0
Kirghizstan	1985	34.0	6.5	182.0	124.0	31.9	9.1	144.0	65.0	98.0	25.0
	1992	40.0	1.9	212.0	117,0	23.6	7.4	135.0	62.0	71.0	23.0
Tajikistan	1985	27.0	3.3	152.0	104.0	25.2	11.4	178.0	35.0	91.0	38.0
	1992	12.0	1.3	111.0	57.0	9.7	8.1	110.0	17.0	17.0	21.0
Turkmenistan	1985	37.0	4.6	168.0	92.0	27.6	8.0	168.0	31.0	111.0	18.0
	1992	30.0	1.5	177.0	72.0	20.0	8.5	170.0	23.0	29.9	31.0
Uzbekistan	1985	28.0	5.0	180.0	170.0	2.8	11.2	177.0	26.0	107.0	31.0
	1992	21.0	2.1	167.0	67.0	12.6	12.4	162.0	23.0	27.9	43.0

Table 8.9. Results of selective questioning of population in 1990 (Inquest of population.., 1991).

Countries	Percentage of people, who believed they enough consumed								
	Bread	Grain products	Potatos	Dairy products	Sugar	Confectionery	Vegetables & fruits	Fish	Meat
Kazakhstan	97.5	81.7	74.9	59.4	27.7	26.1	34.3	26.4	23.7
Kirghizstan	96.6	86.1	74.4	49.4	31.2	44.3	48.7	35.9	16.7
Tajikistan	95.7	75.8	76.9	54.0	25.6	20.7	51.1	22.9	8.0
Turkmenistan	99.1	88.9	71.1	52.0	50.9	39.7	44.6	38.3	24.0
Uzbekistan	99.2	88.1	82.5	43.5	44.7	50.5	47.5	28.0	10.7

Desertification and human health

Under the impact of desertification, the population of areas affected by desertification suffer first of all from bad quality of water. In the zone of ecological disaster, the people consume food contaminated by pesticides and other chemicals used in agriculture. In the Tashauz velayat and other districts of Turkmenistan chemicals are accumulated in storage lakes and water reservoirs. For example, water of the Sarykamysh lake is contaminated by pesticides and other chemicals brought by drainage water. In spite of that, fishing has been continuing in this lake. B. Uraev (1995) found that in October 1995, the daily catch of sazan (a fish from the carp family), pike, pirch, vobla (Caspian roach) and other fishes totaled 24 MT. People also consumed fish from the Khauz-Khan and other storage lakes of Turkmenistan. According to Seiitkurbanov (1991), 0.7 mg of pesticides and other harmful chemicals per kg of fish body were registered in fish caught in these lakes.

A special mission of the World Bank (1995) studied the water supply systems in Turkmenistan where 55% of population lived in rural country. Infant mortality in this country is the highest in CIS (46 per 1,000 born) and the length of life is the shortest (66 years). Ecological conditions were studied in the Tashauz velayat where 892 thousand people lived. The worsening of the ecological conditions caused an increase of respiratory and infectious diseases. People used potable water of bad quality. 10% of families have access to water-lines, 35% used hand pumps and 44.6% consumed water from wells. Soap was not found in 30% of families. The people spent 85% of the earnings to buy food. That testified to high prices of food and the general poverty of the people.

According to a special study of the World Health Organization (1994) the following health problems are common in the zone of if ecological catastrophe around the Aral Sea:

- Diseases of kidneys and liver, arthritis, chronic bronchitis, enteric fever, hepatitis A, the number of which increased during the last 10 years,
- High neonatal and moternal mortality,
- Increase of the number of inborn diseases, children's diseases, spotted fever, tetanus, etc.; their number could be reduced by vaccination,
- Diseases caused by consumption of bad water and environmental pollution, for example diarrhea,
- Strong respiratory diseases, especially by children,
- Anemia, near all women suffering from it, the disease is also common among men and children.

Statistical information on the sickness rate of the people in this region is given in Table 8.10.

Zholdaev (1990) told about the increase of the number of suicides in the zone of ecological catastrophe around the Aral Sea. Many cases of suicides among women and youths were registered in the desertified areas of Karakalpakia. As known, several cases of suicides among young women were common in Central Asia before. Usually they burned themselves with petrol. But suicides among youths is a new tragedy caused by poverty, ecological stress and unemployment.

Table 8.10. Number of sick people in the Aral Sea region, per 100,000 (WHO, 1994).

Regions	Respiratory diseases	Viral hepatitis	Enteric fever and paratyphoids	Intestinal diseases	Renal diseases	Cancer	Anemia	Dysentery
Karakalpakia	1344.9	506.7	6.9	311.8	209.6	79.0	598.8	51.9
Khorezm	685.6	497.4	3.4	785.6	166.9	57.0	237.9	37.7
Tashauz	2189.4	429.9	4.7	387.0	no info.	no info.	4434.5	11.2

More than 90 persons, including 20 children were killed during a flood that occurred in the Fergana valley, Uzbekistan in 1988 (Anonymous, 1988a). Sh. Karimov, the minister of Public Health, reported that most of the people had come to the village of Shakhimardan for holidays. The flood created a dangerous sanitary situation because of the possibility of spreading several diseases, for example, dysentery, enteric fever and salmonnelosis. So the Ministry of Public Health took special preventive measures to provide immunity for local people.

Tedzhen case study

A sociological study was conducted by the Desert Research Institute, Ashkhabad in the frame of the project on Turkmenistan Preparatory Activities to Combat Desertification (Babaev & Kharin, 1994a). The experimental area was located in Tedzhen district, belonging to Akhal velayat. The Karakum Canal and the Tedzhen river are the main water-ways in the area. The Tedzhen is "a lost river". From ancient times to present days its water has been taken for irrigation. The amount of salts in its water was 0.9 – 1,7 g/l.

The ecological degradation in the area under consideration is closely correlated with physical features and human activity. The main factors stimulating the degradation of land are as follows: flat topography, heavy mechanical composition of soils, salinization and water logging, insufficient length of drainage network and overgrazing of rangelands.

The length of the drainage network in oases totaled 28 linear m per hectare, instead of 40–50 linear m recommended by scientists. Irrigation canals, 4,000 km in length were constructed in earth beds, that caused a 50% lost of irrigation water. In general, moderately and severely degraded soils occupied 60% of the irrigated fields.

Physical properties of the soils were bad because of the use of heavy tractors, which pressed and sealed the earth surface. Soil fertility was low because of the absence of crop rotation and manure, which was not used in agriculture. The length of forest protective belt is low. Planted trees and shrubs died under the conditions of saline soils and saline ground waters.

In 1992 the cultivated area totaled 271,165 hectares, of which 42% was occupied by cotton, 29% by grain crops, 20% by lucerne and the rest by vegetables. Productivity of agricultural crops was low, for example the yield of cotton totaled 1.9 MT per hectare.

Population consisted of 19 ethnic groups (Turkmen, Russians, Kazakhs, Persians, Tatars, Uzbeks, Bashkirs, etc.). All these people lived in harmony and no clashes was registered here between them. Distribution of population by age groups is given below (see Table 8.11).

The age structure of population in Tedzhen is common to all Moslem countries of Central Asia. In general, the population younger than 15 years totaled 43%, the population able to work (15 – 60 years) totaled 52%. The number of old people is very low – only 5%. That testifies to the low standard of living in rural communities of Turkmenistan. Most of the working people (27,300 persons) were employed in agriculture. From them, about 50% were herdsmen and the other half were agriculturists.

Potable water in Tedzhen was of bad quality because of the high content of salts, pesticides and other chemicals. The main part of the population consumed water from irrigation canals and wells. Only 55 % of people were provided with water-lines. Unsanitary conditions were the main reason for high mortality, especially that of children. 43 babies per 1,000 died before reaching 1 year of age.

The nutrition of the people is insufficient. The main part of products (meat, vegetables, fruits, melons) they produced themselves. But in general food was poor in calories. According to information from the Tedzhen Sanitary Epidemic Station, children in kindergardens, 3 – 7 years of age got 219 –1241 kg/calories per day instead of 1,870 kg/calories recommended by nutritionists. Young people, 15 –18 years got 615 – 886 kg/calories per day instead of the 3,400 they needed.

16,714 school children learned in 33 schools. 19 school buildings were not suitable for teaching. In 19 schools children learned in two shifts.

Medical service in the etrap could be characterized by the following figures (per 10,000 people):

	Tedzhen etrap	Turkmenistan as a whole
Number of doctors	11.7	36.2
Number of trained nurses	67.5	97.5
Number of beds in hospitals	76.0	114.0

Data on infectious diseases in etrap are given in Table 8.12.

Table 8.11. Distribution of population by age groups in Tedzhen (Babaev & Kharin, 1994a).

Age groups	Population in rural country		Population in Tedzhen town	
	Number	Percentage	Number	Percentage
<5	10643	17.4	6524	16.2
5 – 10	8668	14.1	5304	13.2
10 – 15	7581	12.4	4621	11.6
15 – 20	6438	10.5	3852	9.6
20 – 25	6006	9.8	4262	10.6
25 – 30	5440	8.9	3831	9.5
30 – 35	4320	6.9	3035	7.6
35 – 40	2903	4.7	2064	5.1
40 – 45	1496	2.4	1123	2.8
45 – 50	1386	2.3	1062	2.6
50 – 55	1622	2.6	1270	3.2
55 – 60	1607	2.6	1074	2.7
60 – 65	1208	2.0	852	2.1
65 – 70	810	1.3	550	1.4
>70	1144	1.9	735	1.8
Total	**61272**	**100.0**	**40159**	**100.0**

Table 8.12. Infectious diseases in the Tedzhen etrap (Babaev & Kharin, 1994a)

Diseases	1991		1992		1993	
	Number of sick people	Died	Number of sick people	Died	Number of sick people	Died
Enteric fever	44	2	49	None	18	1
Dysentery	218	3	250	2	223	2
Hepatitis	342	5	121	2	204	2
Respiratory diseases	240	4	160	None	773	None
Measles	1	None	10	None	82	None
Tuberculosis	28	8	33	7	26	5

8.4. Enormous economic and social damage caused by desertification

United Nations statistics on the extent and cost of desertification give only global figures and are based on rough estimation of desertification damage. Inaccurate statistics can lead to a false appreciation of the problem, as Stiles (1995) said. In connection with that wrong decisions can be made by planning measures on a national level. Many experts have tried to estimate desertification damage in terms of money. For example, by estimation of FAO, 21 African countries suffered from desertification in 1984 – 1985 needed food aid of cereals at a cost 3.5 billion US dollars (Tolba, 1985).

Nan-Ting Chon & Dregne (1993) have given several principal propositions on cost/benefit analysis of desertification which are of great interest. First of all they recommended to estimate the damage of desertification for each process separately. For example on-site damage of wind erosion includes nutrient removal, and off-site damage includes air pollution. In several case on-site damage is documented, that makes easier its estimation. But it is very difficult to estimate the social cost of desertification, because it can be intangeble. As the authors of this study emphasize the connection between causes and effect of desertification are not always well understood.

Would it be possible to transform desert into "flourishing land"? Most of countries affected by desertification are poor developing countries. Before spending money for rehabilitation of desert land, decision makers and planners must know the real ecological situation. Ecological constraints always exist and they must be overcome in the operation of any project (Figure 8.1).

Rehabilitation of degarded land means turning the process of desertification backwards. But we don't know if the process of desertification is reversible. According to Floret & Pontanier (1982), reversibility is the ability of a degraded ecosystem to recover its conditions after 25 years of total protection. We may suppose that in several cases the process of desertification is irreversible. In this case two alternatives can be selected: (1) conservation of degraded land, by the lack of money, (2) creation of new anthropogenic ecosystems differing from that which existed before, if it is economically possible.

From this point of view it would be very interesting to assess the real situation in different geographical regions. Such an attempt was made in Turkmenistan by Svintsov (1988). He has divided the territory of the Karalum desert into three categories:

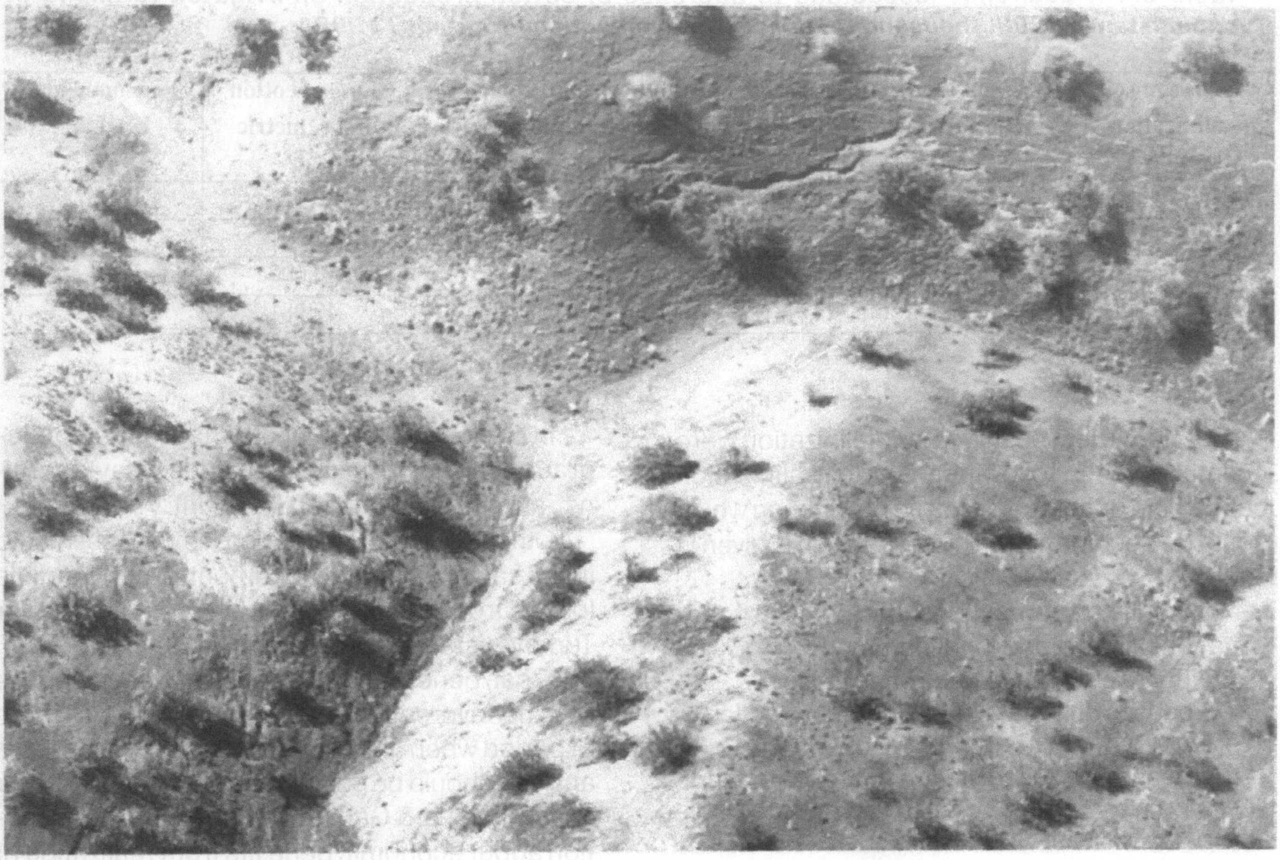

Figure 8.1. Pistachio-tree in the Badkhyz reserve, Turkmenistan where the measures on protection and regeneration were taken. Large scale aerial photo.

1. Areas suitable for afforestation of sand dunes (1,519 mln. ha, or 11%);
2. Areas in which productive rangeland could be established (2,584 mln. ha, or 20%);
3. Low productive sparse woodland not suitable for improvement (9.219 mln. he, or 69%).

That means, only 31% of the territory of Karakum could be restored in case of desertification. These areas could not only be restored but their productivity could be increased. The rest needs mainly conservation because of ecological constraints (high soil salinity, stony surface, undeveloped soil profile, etc.). In case of desertification these areas should be protected and restored by special measures, for example limitation of movement of transportation vehicles, restriction of animal grazing, etc. Active measures, like sand stabilization should be conducted here only around settlements or near gas pipe lines and other strategic objects.

By preparation of the National Action Program to Combat Desertification in Turkmenistan (1996) the following classification of desertification damage was used. According to this approach direct documented losses (the income foregone) included:

- Loss of animal production on grazing land,
- Loss of agricultural production on irrigated farmland,
- Loss of agricultural production on dry farmland,
- Loss of wood in degraded forests,
- Loss of other forest products,
- Loss of fish production in rivers and lakes contaminated by water conducting from degraded irrigation fields.

Table 8.13. Economic losses due to decrease of cotton production on irrigated farmland of Turkmenistan in 1993, thousand US dollar (National Action Program.., 1996).

Degree of soil salinization	Size of area, hectares	Decrease of cotton production, %	Loss of row cotton, metric tons	Loss of cotton fibre, metric tons	Economic loss, US dollar
Slight	166, 586	15	49,976	16,658	17,057.8
Moderate	314,407	30	188,428	0,628	64,316.4
Severe	76,503	60	91,816	30,605	31,339.5
Total			330,816	47,981	112,713.7

Economic losses due to soil salinization were estimated in the following way. Soil salinization caused the decrease of cotton production which could be documented. This calculation is given in Table 8.13.

Indirect losses included the cost on tehabilitation of degraded land, namely:

- Expenses on rangeland rehabilitation,
- Expenses on rehabilitation of degraded farmland,
- Expenses on rehabilitation of dry farmland,
- Expenses on rehabilitation of degraded forests,
- Expenses on control of wind erosion,
- Expenses on control of water erosion and mud flows.

In Table 8.14 the cost of rehabilitation of degraded farmland is given on each velayt (adminstrative region) and for the country as a whole.

Other direct and indirect losses were estimated in the same way. Total documented and tangible losses from desertification were estimated for Turkmenistan. They are given in Table 8.15.

Cost/benefit analysis of desertification on the national level could give information on the length of period when all expenses will be justified. This analysis should be conducted with great care. For example, Zha & Gao (1997) published information about economic bene fits from realization of one of the greatest projects of present days, the so called "Green Great Wall" project, embracing 4,069 mln. km^2 in North China. According to this

Table 8.14. Cost of works on rehabilitation of degraded irrigated farmlands in Turkmenistan, thousand USD (National Action Program.., 1996).

Velayat	Construction of the horizontal drainage	Leaching the soils	Total
Akhal	2,309.6	16,414.3	18,723.9
Balkan	76.9	6,064.7	6,146.6
Tashauz	1,809.6	12,080.9	13,890.5
Lebap	592.2	2,395.2	2,987.4
Mary	2,306.0	20,678.5	22,984.5
Total	7,094.3	57,633.6	64,727.9

information, direct benefits from realization of this project will total by 2025 about 25,55 billion yuan. Direct benefit will be 27 times greater then the initial investment. The indirect benefit was valued at 44,6 billion yuan. But the authors of this publication didn't give exact calculation of all debit and credit items.

In our opinion, it is better to calculate land rehabilitation expenses by each process separately. An example can be given on Turkmenistan, where a technique was developed on rangeland improvement by planting forest belts. The cost of these works is given below (See Table 8.16).

Using the same approach, calculations can be made for other types of land degradation. Experts of the Desert Research Institute in Turkmenistan developed a project proposal on melioration of saline soils in Tagry etrap (administrative district).

According to the calculations, the cost of of the project totaled $11.7 mln. That included the meliorative works on an area of 50,000 ha. By implementation of this project, its expenses could be justified in 10 – 15 years (National Action Program.., 1996).

Cost/benefit analysis of desertification has shown that a special methodological approach should be applied by estimation of damage caused by land degradation. Desertication damage includes: direct losses (income foregone), cost of land rehabilitation and social losses which are not documented and often are not tangible. Direct losses and cost of land rehabilitation can be estimated in monetary terms.

Table 8.15. Economic losses from desertification in Turkmenistan, thousand USD (National Action Program.., 1996).

Denomination	Losses, thousand $
Direct losses (income foregone)	
A. Losses of animal production	160.6
B. Losses of agricultural production	112,713.7
Total A + B	112,874.3
Indirect losses	
C. Expenses on rehabilitation of degraded rangeland	156,780.0
D. Expenses on rehabilitation of degraded farmland	64,727.9
E. Expenses on forest regeneration	768.9
F. Expenses on sand stabilization and afforestation	11,727.7
Total C + D + E + F	234,001.5
Total A + B + C + D + E + F	346,875.8

Table 8.16. Cost/benefit analysis of rangeland improvement in Tuerkmenistan, per 100 ha (Svintsov, 1988).

Criteria	Interval between forest belts, m		
	40	30	20
Capital investments, rouble	1141	1344	1606
Benefit: additional forage, MT	6.6	7.8	9.2
Additional wood, m^3	190	200	210
Pure income, rouble: livestock production	82	96	120
Wood	95	100	105
Total income, rouble	177	196	225
Number of years during which all expenses will be justified	6.5	6.9	7.1

Note: All expenses are given in prices of 1988.

The system of sanitary control of agricultural pests which functioned during the Soviet period was destroyed after 1991. New independent countries were not ready to control locust and other pests, especially in Kazakhstan. According to published information (Anonymous,

1999b), some 9 mln. ha of agricultural crops in this country were damaged in 1999 by locust, and the yield was fully destroyed. The government of Kazakhstan applied to the Bank of Development of Asia for a loan of $10 mln. In the budget of the country more than $11.3 mln. would be spent for pest control. But the financial situation in the country is critical and there was no hope that this money would be invested in this project.

8.5. Conclusion

Disintegration of the USSR caused the collapse of the economy of the former Soviet Republics of Central Asia. New political regimes were formed in the region. Political instability, poverty, corruption and mass movement of people within and beyond the region aggravated the ecological situation.

Scarcity of water resources in Central Asia is one of the urgent AEP. By estimate of experts, the projected water demand in the Aral Sea basin totals 127 mln. km^3. Deficiency of water resources will increase 30 – 40% in the beginning of the 21st century.

Desertification being one of the AEP hits not only the desert environment but also the economic and social life. Economic losses from desertification include direct losses (income foregone) and indirect losses (expenses on rehabilitation of degraded lands). By rough estimate, in Turkmenistan alone, the total losses from desertification were $346,875.8 per annum.

Social losses can't be expressed in terms of money, but they also are great. The population of the areas affected by desertification suffers first of all from bad quality of water. According to a special study, many health problems are common in these areas, and the length of life of people is reduced. Small children and old people are affected more than other sectors of the population. Medical service is very bad because of a shortage of personnel and economic difficulties.

CHAPTER 9

CONVENTION TO COMBAT DESERTIFICATION – A GLEAM OF HOPE AT THE END OF THE TUNNEL

It is a long lane that has no turning.

English proverb.

9.1. A concept for rehabilitation of degraded desert lands

Before taking measures on desertification control, ecologists must give information to planners about the degree of desertification and about the ecological cons traints of lands to be improved. Would it be possible to transform desert into "a flourishing land", as apologists of nature remaking of Stalin's epoch demanded? We must be realistic and we must attack more limited tasks – rehabilitation of desertified lands. We may suppose that in several cases the process of desertification is irreversible. According to Floret & Potanier (1982), reversibility is the ability of a degraded ecosystem to recover its conditions after 25 years of total protection. We may suppose that in several cases the process of desertification is irreversible. In this case ameliorative measures should be aimed at creation of a new anthropogenic ecosystem differing from that existed before.

The conceptual approach for planning measures on land melioration by different degree of desertification is given below (see Table 9.1). Dividing the degraded lands into three classes corresponds to classification of measures on land melioration proposed by UNEP.

We have calculated the size of agricultural lands needing melioration in Central Asia (Table 9.2). It should be noted here that melioration of irrigated farmlands is the most urgent problem in the region because the population needs food.

Scientists of Turkmenistan have developed several technique on improvement of desert rangelands in the Karakum desert and in the foothill country (Nechaeva & Prikhod'ko, 1966; Mukhammedov, 1979; Svintsov, 1988). By proper management, all expenses can be justified in 5 – 10 years.

Melioration of saline farmlands is the most expensive and labour-consuming activity. It includes: reconstruction of existing drainage network, construction of new irrigation canals, development of crop rotation system, etc. Experts of the Desert Research Institute developed a project proposal on melioration of saline soils in Tagty etrap of Turkmenistan (National Action Program.., 1996). According to calculations, the cost of the project totaled $11.7 mln. That included the ameliorative works on an area of 50,000 hectares, where 50% of the area had been moderately affected by desertification, and 72% had a shallow ground water table. Implementation of this project could be justified in 10 – 15 years.

It was also known that forest belts protect fields from sukhovei (dry wind) and stimulate the growth of agricultural crops. But agronomists

Table 9.1. Measures on land melioration depending on the degree of land degradation (Kharin, 1996a)

Concept	Degree of land degradation		
	From non degraded to slightly degraded	Moderately degraded	Severely degraded
Preventing measures	Conservation and rational exploitation		
Correcting measures		Rational exploitation combined with partial melioration	
Measures on full rehabilitation of degraded land			Melioration of the whole area
Supporting measures	Capacity building, training, research, etc.		

Table 9.2. Measures on land melioration in Central Asia, in million hectares/percentage.

Land use classes	Conservation or rational exploitation	Rational exploitation combined with partial melioration	Melioration of the whole area	Total
Forests	22.561/69	10.438/31	0.672/2	33.671/100
Rangelands	16.997/69	7.145/29	0.492/2	24.634/100
Irrigated farmlands	0.867/10	7.025/81	0.780/9	8.672/100

sometimes raised objections against introduction of forest belts because the belts occupied a part of land which was used for cultivation of a main crop, and also they used irrigation water. But the data of Table 9.3 compiled by Russian scientists from the former Department of Forestry of the USSR (Isaev, 1991), show that afforestation is prefrable. So, protective forest belts can be cultivated in oases and around oases of Central Asia.

Table 9.3. Increase of cotton yield on the fields bordered with protective forest belts in Uzbekistan (Isaev, 1991).

Activity of wind	The rate of yield increase		Yield increase taking into consideration the areas of forest belts	
	kg per ha	%	kg per ha	%
Strong	420	15	390	14
Moderate	380	12	300	10
Slight	230	7	160	5

The era of nature remakers in central Asia has come again. S. Niyazov, President of Turkmenistan issued a special decree about implementation of the project "The Green Belt" (Mikhailov & Smol'nikov, 1999). This program would include creation of protective belts around all big cities and administrative centers of the country. Green plantings would include 6 rows, each 36 m in width. Young trees, 2 – 3 years old, grown in special nurseries would be planted in these belts. Several tree species, like pine, maple, juniper and others would be used in this project. According to the decree, a special joint stock company "The Green Belt" would be set up. Niyazov called on all inhabitants of the country to support this project. The authors of this ambitious project believed that green plantings could ameliorate the microclimate and protect the population centers from hot winds blowing from the desert.

9.2. Regeneration of forests

A special team of Russian scientists in the Department of Forestry of the former USSR developed a program for forest regeneration in Central Asian republics, except Kazakhstan (Isaev, 1991). Probably, new independent countries will not follow these recommendations, but it will be so much the worse for these countries. Official statistics doesn't give figures about the reduction of forest area during the last decade, but the share of productive forests (stocked forests) is very low, only 51%. The rest are understocked forests which can't fulfill the protective functions. The size of forest plantations of four countries is given in Table 9.4.

But the measures on forest regeneration were not successful because of low ability of young trees to take roots under severe environmental conditions.

Table 9.4. The area of forest plantations in four countries of Central Asia (Isaev, 1991)

Countries	Area of forest plantations		Annual size of forest planting and seeding	
	Total area, thousand hectares	Per 100 hectares of forest land, ha	Total, thousand hectares	Per 100 hectares of forest land, hectares
Kirghizstan	66.8	7.12	2.30	0.24
Tajikistan	50.3	7.01	3.14	0.44
Turkmenistan	144.0	1.72	46.67	0.56
Uzbekistan	461.0	12.60	25.48	0.71
Total	**723.5**	**5.28**	**77.59**	**0.57**

If we believe the official data, the following areas of forest plantations survived: 23% during 1961 – 1982 and 57.7% during 1983 – 1987. In spite of that the authors of recommendations mentioned above said that cultivation of forest plantations is economically profitable. The coefficient of economic efficiency for growing forest plantations was: 1.89 – for *Picea schrenkia-na,* 6.7 – for archa, 1.15 – for *Juglans regia,* 3.77 – for *Populus* species and 2.53 – for saxaul species. Wind erosion protective belts were also effective. The coefficient of economic efficiency was: 1.37 by severe wind erosion, 1.95 by medium wind erosion and 2.2 by slight wind erosion.

The following recommendations are given for further regeneration of forests:

- Reconstruction of mountain, tugai and desert forests aimed at the increase of their protective functions and production of additional wood,
- Growing new productive forest stands of valuable tree, especially that of coniferous forests,
- Growing new plantation of wild fruit and nut trees (*Juglans regia, Pistacia vera, Amygdalus communis* and apricot trees),
- Creation of new protective forest belts on mountain slopes, around oases and within irrigation zone, especially in the areas of cotton cultivation,
- Creation of protective forest belts near irrigation canals, roads and gas-mains.

In general, the forests of four countries of Central Asia need measures of forest regeneration in the area of 7,239.5 thousands hectares. From this area, 22% can be regenerated by natural processes, the rest needs artificial measures.

9.3. New approaches and arrangements

The United Nations Conference on Desertification held in Nairobi in 1977 adopted "The Action Plan to Combat Desertification". The Plan was not implemented but small progress in desertification control was achieved at a local level. The Earth Summit held in Rio in 1992 was the next milestone in international efforts to combat desertification. This summit adopted the concept of sustainable development and recommended that the General Assembly of UN develop "The Convention to Combat Desertification" (CCD).

Sustainable development

CCD promotes several new approaches which should be adopted by the world community. The concept of sustainable development should be mentioned here first of all. According to FAO's definition, "sustainable development is the management of the resource base, and the orientation of technological change in such a manner as to ensure the attainment and continued satisfaction of human needs for present and future generations". This concept is aimed at conservation of land, water and forage resources, it will also contribute to conservation of genopools of plants and animals.

It should be noted here that sustainable development is not a new concept. A concept closely correlated with sustainable development was known in forestry of Germany in the 18th century. That was a concept of "sustained – yield forest management". The term can be used in two sense: as continuity of growth or continuity of harvest. Sustained – yield management , as the term is most accurately and commonly employed, means continuity of harvest. The concept was based on the assumption that exploitation of forests, mainly forest cutting should be continuous and equal *ad infinitum* in the limits of forest districts. Later the concept was adopted by the foresters of Russia; it was included in 1845 into the "Manual of forest management" (Orlov, 1927).

We can give an example of application of this concept in very simple form in forests of Siberia. At the end of the 8th century, many Russian business men, so called "kupets" (merchants) took

According to this classification, vast areas of forests in Siberia and in other regions of the USSR were considered as the forests of the third group. As a result of this approach, rapacious exploitation of forests began. Desert woodland and mountain forests of Central Asia were classified as the forests of the first group. But, because of shortage of wood in the region these regulations were not always observed.

In range management, this concept can be realized by application of intermittent grazing, rotation grazing and some other systems which can provide regeneration of forage plants. But in Central Asia these systems were never introduced in spite of some successful experiments conducted during the Soviet period.

But we have very good examples when the concept of sustainable development can be applied successfully. The preconditions for its implementations are given by the decree of Niyazov. According to it, all people in Turkmenistan are freely provided with gas, electric power and salt (Babaev & Kharin, 1994a). So, the people living in the desert will not cut saxaul and other trees and shrubs for fuel. At least, a natural regeneration of woodland would be provided.

The countries of Central Asia adopted "The Nukus Declaration"in September 1995. In this Declaration they approved the concept of sustainable development and gave obliged themselves to join International Conventions, in particular CCD, Convention on Biodiversity Conservation and Convention on Climate Change.They also expressed their full support of regional organizations created in Central Asia.

However, the practical realization of this concept in the region met many difficulties. The main bottle – necks included political, economic and ecological factors. Independence and establishing new political regimes could be named as political factors. Political instability and civil war in Tajikistan destroyed the economy of this country and caused economic damage totaling billions of roubles.

The sovereign countries of Central Asia have inherited the strictly centralized, so called "planned economy" from the former USSR. This economic system resulted in uneven economic development of the former Soviet republics, non-effective utilization of the natural resources and labour potential. Serious disadvantages in the economy of these republics were non-integrated development, bad management and wrong distribution of industrial enterprises, especially those which stayed under the control of Moscow. Some branches were based on utilization of row material delivered from remote regions. Ecological aspects were not taken into consideration by industrial development.

Economic reforms in Central Asian countries have made a small progress. Only Uzbekistan is ahead of other countries. In 1994, 54,000 enterprises were privatized. In this country the portion of the non-governmental section was 40%, while in Turkmenistan only 5% (Babaev & Kharin, 1994c).

Agricultural reforms started in Tajikistan in spite of civil war. According to land tenure reform, each collective farmer could get personal use a plot of land, up to 50 hectares in size. In several cases farmers dragged local administrators from their offices and demanded the implementation of the decree on the allotment of land *in situ* (Gusseinov, 1996).

Bad management and corruption were also constraints on sustainable development. "Pripiska" so common during the Soviet period proved to be actual in the post communist society. Turkmenian State Television reported on 16 March 1996 that Niyazov, president of the country, visited two velayats where local administration had falsified reports on production of wheat in 1995. Some of them were dismissed from their posts. Agabaev (1996) cited the words of Niyazov who said: "You administrators didn't cope with your tasks. You did that deliberately... Who forced you to commit this fraud? You betrayed not only my trust, you deceived the State and the people. That was a grave crime. You must atone for your guilt with honest work".

What is better – top-down approach or bottom-up approach?

CCD proposes to replace "top-down" approach" by "bottom-up approach". That means more active participation of local communities in anti-desertification activities. Ordinary people in rural communities always participated in all measures of land melioration but they used to do that only by orders given "from above", from the higher authority. Any initiatives from the grass-roots were usually suppressed. We must take into consideration specific features in life of Moslem communities in Central Asia. Traditionally, people always stayed under the yoke of a khan, beg, chairman of kolkhoz or raikom secretary. Many generations of people were raised under the fear before the powers that be. Their life and property were in hands of these mighty princelings. Ecclesiastics, like mullahs and imams, usually supported this system. In the mind of ordinary people, these rulers were divided into two categories: good and bad; a good raikom secretary fleeced the people to a smaller extent than a bad raikom secretary. The situation really didn't change after proclaiming independence. New administrators (named "khya-kims") replaced raikom secretaries but their attitude towards the local people was the same.

A very interesting sociological study was conducted in Uzbekistan by Kakhkharov (1993). He studied the social activity of local people. All people interviewed were divided into three groups depending on the degree of their social activity:

1. Little active (10%),
2. Moderately active (78.1%),
3. High active (11.9%).

The people of the first and the second groups said that only the local administration would improve social and ecological conditions in rural areas. They were sure that their own role in decision making was very low. So they would wait for a long time...

Above that, there is an other problem dealing with management of local resources. Local people in any political system try to satisfy their needs using local resources. At first sight, this approach was well justified. But in many cases that contradicted the strategy of nature conservation proclaimed by national government.

During the Soviet period many conflicts with local people arose in the Turkmen SSR by widening the areas of reserves and national parks. The government of this republic tried many times to enlarge the size of the protected areas. The Suynt-Khasardag Reserve located in the

Kopetdag mountains was one of the objects of this activity. Many rare and endangered plant species in this area were on the verge of disappearance. Local people from the tribe Nukhur were deprived of grazing lands by these regulations. So the Nukhur people "declared war" against rangers of the reserve. They also sent complaints to Moscow, demonstrated before the residence of the Central Committee of the Communist Party in Ashkhabad, etc. At last a new government of independent Turkmenistan reduced the size of this reserve to such an extent that the reserve couldn't fulfil its functions...

The problem needs a special solution based on compromise between two hostile parties – local people and government.

Capacity building

Capacity building includes several measures, in particular strengthening scientific potential, improvement of training cadre, especially at low level, popularizing new technologies among local population, development of environmentally safe technologies in agriculture, education of specialists and officials, etc.

A well developed infrastructure of training cadres was created in Central Asia during the Soviet period. This system included: educational and research institutions, technical colleges and professional schools. The central government in Moscow controlled this system. Diplomas of pro- fessor, doctor and candidate were given by The High Certifying Commission (VAK). This agency made a final decision on awarding scientific degrees.

This system was fully destroyed in 1991. So new independent countries met difficulties in creation of a new system of training cadre. The main difficulties were the following:

1. Mass migration of Russian scientists and specialists. As known the Russians were bearers of high technical potential. Now it is clear that this potential will be lost in the nearest future.
2. Training cadre in local languages. During the Soviet period all text books and manuals were printed in Russian. Local specialists didn't master scientific terminology in their own languages. They mainly used a mother tongue as a "market language". Preparation and printing of text books in local languages need time and money.
3. New independent countries try to replace the former corrupted educational system by a new (more corrupted?) system. New VAKs were set up in all countries of Central Asia. For example, in Turkmenistan a new system of selection of young people was introduced. Special commissions consisting of school teachers, local administrators and university professors selected candidates for university education. This system opened a new way for corruption and bribes.
4. The craving for knowledge seized the people of sovereign Uzbekistan!.. Entrance examinations to universities were abolished, the new Uzbek VAK granted scientific titles to those who wished. During the Soviet period, the central VAK in Moscow awarded the degree of doctor to 15 – 20 Uzbek scientists annually. In 3 years of independence the Uzbek VAK awarded the degree of doctor to 700 candidates. At last, president Karimov closed down the Uzbek VAK (Vinogradov, 1996). Now those who wish to get a scientific title have to go to neighboring countries. But in the neighboring country, in Turkmenistan VAK was also closed two years later!..
5. The whole system of scientific investigations needs restructuring. Scientific institutions and universities lost many scientists, especially Russians. Young people don't want to undertake post graduate studentships because of low wages. Local libraries don't receive publications in Russian and other European languages. Scientific equipment being very poor during the Soviet period is out of operation now, etc.

Ecological education, being so incomplete during the Soviet period, is now suffering from depression. For example, A. Agaev, a school teacher of the Magtymguly etrap of Turkmenistan told the following. Ecological problems were discussed in lessons on botany, biology and natural history. But schools were not provided with visual aids and booklets. O. Kulieva, the Second Secretary of the Democratic Party of the same etrap said the following:"Ecological education in the etraps is at a low level. Articles published in newspapers are not interesting, and the people don't read them. We need good lecturers who will distribute ecological knowledge among the plain people, among women and children" (Babaev & Kharin, 1994a).

Poor ecological education of people, especially that of children, could cause a real damage to environment. For example, fire destroyed vegetation on an area of 720 ha in the Syunt-Khasardag Reserve of Turkmenistan (Ustaev, 1995a). The fire arose from a schoolboy who made a bonfire. His parents paid a considerable fine. During the same year two big fires occurred in the area under consideration. The author of this information said that fires were common each year in the reserve and in the bordering areas.

Participation of NGOs in capacity building is very important. T. Mollaev, a member of the Ecological Club in Tashauz velayat of Turkmenistan proposed a project on training agronomists, team leaders of collective farms and farmers in agrochemisrty (Anonymous, 1994b). Application of mineral fertilizers in the velayat was uncontrolled. Not a single specialist in agro-chemistry was on the staff of collective farms or the agricultural administration of the velayat under consideration.. Nitrous fertilizers were used on cotton fields without any control. But as known, these fertilizers should be used in combination with phosphates and potassium. T. Mollaev began training people by financial support of foreign donors.

Promotion of new sources of income in agriculture

People in local communities should be active participants in implementation of any program to combat desertification. The situation in rural areas of Turkmenistan is the following (Babaev & Kharin, 1994a). A special investigation was conducted in Bakharden etrap of Turkmenistan. In this area the income of local people consists of two sources: wages and non-fixed income. Mean wages of people working in kolkhoz and sovkhoz in 1992 totaled 4,056 roubles per month (about $ 4). The main sources of non-fixed income included:

- Payment in kind for work in agriculture,
- Material aid provided by kolkhoz to poor families (families devoid of a bread winner, unmarried mothers, families of elder people),
- Material aid for newly-weds (they usually received 2 sheep and some products for wedding),
- Income from domestic crafts,
- Income from personal plots, etc.

G.Orazov, 52, a leas eholder from kolkhoz "Durun" told the following. He took on lease 4,0 hectares of land and cultivated cotton. In 1992 he earned 21,899 roubles per month (US $ 22). He also had a personal plot of land where he cultivated vegetables for personal consumption. He also had one cow and six sheep. Forage for livestock is provided by the kolkhoz at a low price. In spite of low fixed income, 70% of families in this kolkhoz had personal cars or motorcycles and 99% of them had TV sets and refrigerators.

Many small enterprises were functioning in etrap. They processed agricultural products, and produced dried melons, currants, woolen yarn, etc. Two cooperatives grew mushrooms, some of them were engaged in agriculture, fishing, growing medical plants, etc. Sewing workshops were also functioning in a kolkhoz in which 30 local women were employed.

The inquiry of the population was also conducted in two other districts of Turkmenistan. Summarizing this inquiry, the following measures should be named to increase income:

- Organization of cooperatives on purchase of agricultural products,
- Rehabilitation of the forgotten traditional technology of milking sheep. In the past local women milked sheep during two spring months. "Sargan" (a kind of cheese) was produced from the first milk of sheep after the birth of lambs. Butter was also produced from sour milk. Fatless milk was used for producing several products and in folk medicines. Women used it for washing hair.

- Organization of special centers for hiring agricultura l machines by leaseholders and private owners of land.

9.4. National Action Program to Combat Desertification (NAPCD) in Turkmenistan

Measures to combat desertification at national level

CCD is to be implemented through the NAPCD at a national level. Only two countries of Central Asia – Kazakhstan and Turkmenistan have prepared NAPCDs. CCD lays several general requirements for the NAPCD and voluntary elements. Each affected country could include at least some of these elements in NAPCD. NAPCD in Turkmenistan was based on previous experience and preparatory activities to combat desertification (Babaev & Kharin, 1994a). It was developed in the Desert Research Institute under the guidance of A.G. Babaev and N.G.Kharin (National Action Program.., 1996). Several scientists participated in preparation of this program, in particular: M. Aranbaev, A. Batyrov, M. Kolodin, O. Kurbanov, G. Mukhammedov, M. Nurberdyev, A. Ovezliev, N. Orlovsky, K. Redzhepbaev, I. Svintsov, P. Esenov. It was "a program maximum" which could be realized through many projects. The NAPCD consisted of the following sections:

1. Physical environment and resources,
2. Socio-economic development,
3. Human-induced desertification,
4. Strategy and plan of action,
5. Program of high priority to combat desertification.

Strategy and plan of action included several recommendations, the most important of them are given below.

Development of National Monitoring System

Decision makers and experts need operative information on land degradation for planning measures on desertification control. CCD insures the systematic observations of land degradation and creation of early warning system. The National Center on Desertification Monitoring (NCDM) will be set up in Turkmenistan. NCDM will collect information submitted from velayats (provinces) and from etraps (districts). Information on desertification status and on socio–economic conditions will be published in a special "Bulletin of Early Warning". This bulletin will be distributed among local population. The minimum data sets should include:

- Climate variables (albedo, rainfall, air temperature, wind velocity, dust storms),
- Soil and water variables (ground water, major surface water, wind erosion, water erosion, soil salinization, waterlogging, soil compaction and sealing),
- Vegetation variables (plant species composition, woody biomass, disappeared plant and animal species),
- Land use variables (land use system, land tenure system, changes in land use, structure of sown areas, yield of major staple crops, composition and number of livestock, animal pressure),
- Socio-economic variables (human population changes, seasonal and annual human migration, infant and adult mortality rates, length of life, human disease status, per capita income distribution, sources of income, market prices for key food stuffs, energy availability and prices).

A small scale desertification map will be annually compiled on the basis of NOAA/AVHRR data. Desertification monitoring should include three chronological periods: retro-monitoring, assessment of present status and forecast for future.

Strategy of utilization of water resources

Water is one of the most important resources in the desert. The share of water resources used in agriculture totaled 92% in 1992. Deficiency of water resources in dry years could reach 1 – 2 km^3 per year. Irrigation rate totaled 1,700 m^3 of water per hectare per year, not exceeding the recommendations of scientists. Efficiency of water use in agriculture didn't exceed 0.60. Several measures were planned to improve the utilization of water resources, in particular: technical improvement of irrigation systems, recon struction and revetment of the main irrigation canals, introduction of a new technique of irrigation etc. After the implementation of the program, the present irrigation system will be reconstructed and 9,000 km of concrete chutes will be constructed. New conduits, 2,500 km in length, will be constructed inside the farm irrigation network. Use of irrigation water will be decreased and annual saving of irrigation water in agriculture would reach 3.5 – 4.0 km³.

Conservation of biodiversity

Conservation of biodiversity is very important for sustainable development. Loss from natural ecosystems of different plant and animal species results as a rule in reduction of biological productivity. The quantity and quality of forage on desert rangelands falls. Many plant and animal species should be conserved as a gene pool for future selection of new varieties to the benefit of people. The following measures were planned for conservation of biodiversity:

- Conducting measures on protection of rare and endangered species *in situ* and *ex situ*.
- Expansion of the areas under reserves and organization of new reserves, in particular: the Central Karakum reserve, 200,000 ha in size and the Balkan reserve, 30,000 ha.
- Study of rare and endangered species and their habitat,
- Organization of new protective areas for conservation of rare and endangered species.

The main areas of conservation of plant and animal species are given in Table 9.5.

List of plants protected in these reserves is given below (Table 9.6). Protected plant species are marked by symbol "+".

The system of land use and nature conservation in protected areas will be developed (closed area, experimental sites, teaching sites, recreation sites, etc.). The complete realization of this program will provide conservation of genopool of all plant and animal species in Turkmenistan.

Special measures will be taken for conservation of rare plant species. In particular, new protected areas will be created for conservation of the following plant species: *Pistacia badkhysi, Pyrus turcomanica, Dendrostellera turcomanica* and *Malus silvestri*. The following plant species will be re-introduced: *Mandragora turcomanica, Acantholimon kjurendaghi*, and *Cleome turcmena* to Syunt-Khasardag reserve; *Pteropyrum auchcri* to the Meana-Chacha protection area, *Salsola chivensis* to Kaplankyr reserve and to the north coast of the Karabogazgol Gulf. New protected area will be created for *Ficus afghanistanica* in canyon Pelengovaly, near the source of the Kashan. *Vicia fedschenkovana* and *Alcea sycophylla* will be introduced to Syunt-Khasardag reserve. Measures will be taken on res-

Table 9.5. The number of plant and animal species to be protected in the reserves of Turkmenistan (National Action Program.., 1996).

Names of reserves	Area, thousand hectares	Geographical location	Number of protected species	
			Plants	Animals
Repetek	36.4	Southeast Karakum	-	14
Khazar	262.8	Shoal of the Caspian Sea	-	27
Badkhyz	87.7	Interfluve of the Kushka and the Tedzhen	7	33
Kopetdag	49.4	Central Kopetdag Mountains	14	37
Synt-Khasardag	29.7	Southeast Kopetdag Mountains	30	31
Kaplankyr	282.6	North Karakum	2	11
Kugitang	27.1	Gissar Mountains	10	7
Amudarya	50.5	Amydarya valley	2	16

toration of rare and endangered plant species *ex-situ*.

Several measures will be also taken for protection of many animals, in particular: *Equus hemionus* (kulan), *Gasella subguttosa* (dzheiran), *Capra acgagrus*, *Cervus elaphus*, *Felis caracal* (karakal), *Hyaena hyaena*, *Lutra lutra*, *Ovis ammon*, *Panthera pardus* and several other species. More than 30 reptiles also need protection, first of all cobra (*Naja oxiana*), *Erix elegans*, *Telescopus rhynopoma*, *Varanus griseus* and others.

For conservation of many animals new protected areas will be created and the size of existing reserves will be enlarged. Special investigations will be started on restoration of rare and endangered plant and animal species.

Melioration of irrigated farm lands

At present, melioration of irrigated farm land is considered as a leading factor of agricultural development which could solve the problem of food production. Irrigation agriculture may give two harvests during one year, taking into consideration abundant thermal resources of the country. It is equal to doubling size of sowing areas in agriculture. Measures on melioration of irrigated farm lands are given in Table 9.7.

A new drainage network will be constructed on an area of 84,400 ha. The repairing of the present drainage network will be accomplished on an area of 96,500 hectares. On saline irrigated soils, the minimum length of drainage network must be 54 m per ha. Complex reconstruction of the irrigation network will be carried out on the area of 476,000 ha.

Range management and improvement

Range management and improvement is in the first rank of importance in the program of desertification control. Desert rangelands are provided with watering points irregularly. Tracts better provided with water are overgrazed. Partial seeding of shrubs on foothill rangelands can improve the productivity of rangelands, and the number of livestock in these areas can increase by 300,000 head. Seeding of shrubs and semishrubs will be conducted on poor stabilized sand dunes and barkhans in the Karakum desert. The main measures of rangeland improvement are given in Table 9.8.

Conservation of forests

According to official data, the area of stocked forests in Turkmenistan reduced by 994,000 ha during 1983 – 1988. Establishment of protective belts and forest tracts can improve this situation. The program envisages conducting the following measures (ha):

Partial tree and shrub seeding in sandy areas	300,000
Seeding of *Pistacia* trees	4,000
Planting of *Juniperus* trees	850
Green plantations in populated areas	1,250
Fire brakes	1,000

Stabilization of moving sands

Moving sand dunes in Turkmenistan occupy about 7% of its territory. These areas have been expanding from year to year because if intensive human activities in the desert. Moving sands cause great damage to the national economy. The following three main types of moving sands are identified in the country (National Action Program.., 1996):

1. Sands located within oases and around oases.
2. Sands of technogenic origin located near engineering structures.
3. Sand dunes located around desert wells and watering points.

Sands of the first type mainly need afforestation and in some cases a preliminary mechanical stabilization should be carried out. The sands of the second and the third types are less favorable for direct afforestation. They need a preliminary mechanical or chemical stablilization (Figure 9.1). The sands of west Turkmenistan and Transunguz Karakum have severe obstacles for afforestation: scarcity of precipitation, saline soils and movement of sand dunes during the whole year. The physical conditions of central and south east Karakum are more favorable because of higher precipitation and more stable sand dunes. The program recommends the following works on sand stabilization (ha):

Table 9.6. The list of plant species protected in the reserves of Turkmenistan (National Action Program.., 1996).

N	Species names	1	2	3	4	5	6	7	8
1	*Allium vavilovii*	-	-	-	+	-	-	-	-
2	*Allochrusa gypsophiloides*	-	-	-	-	-	-	+	-
3	*Amygdalus scoparia*	-	-	-	-	+	-	-	-
4	*Asparagus brachiphyllus*	-	-	-	-	-	-	-	+
5	*Astragalus bobrovii*	-	-	-	-	-	-	+	-
6	*A. kushkensis*	-	-	+	-	-	-	-	-
7	*Atraphaxis badghysi*	-	-	+	-	-	-	-	-
8	*Atropa komarovii*	-	-	-	-	+	-	-	-
9	*Cleome turcmena*	-	-	-	-	+	-	-	-
10	*Colchicum bifolium*	-	-	-	+	-	-	-	-
11	*Cousinia badghysi*	-	-	+	-	-	-	-	-
12	*Crucianella sintenisii*	-	-	-	-	+	-	-	-
13	*Dactylorrhiza flavescens*	-	-	-	-	+	-	-	-
14	*D. umbrosa*	-	-	-	-	+	-	-	-
15	*Epipactis turcomanica*	-	-	-	-	+	-	-	-
16	*E. veratrifolia*	-	-	-	-	+	-	-	-
17	*Eulophia turkestanica*	-	-	-	-	-	-	-	+
18	*Ferula turcomanica*	-	-	-	+	-	-	-	-
19	*Ficus carica*	-	-	-	-	+	-	-	-
20	*Fritillaria raddecana*	-	-	-	-	+	-	-	-
21	*Gymnarrhena micrantha*	-	-	+	-	-	-	-	-
22	*Gypsophila antonianae*	-	-	-	+	-	-	-	-
23	*Hiacintus transcaspica*	-	-	-	+	-	-	-	-
24	*Iris ewbankiana*	-	-	-	-	+	-	-	-
25	*Juglans regia*	-	-	-	-	+	-	-	-
26	*Lagochilus inebrians*	-	-	-	-	-	-	+	-
27	*Lathyrus sphaericus*	-	-	-	-	+	-	-	-
28	*Lepidolopsis turkestanicus*	-	-	-	-	-	-	+	-
29	*Listera ovata*	-	-	-	-	+	-	-	-
30	*Malacocarpus crithmipholius*	-	-	-	-	-	+	-	-
31	*Malus silvestris*	-	-	-	-	+	-	-	-
32	*Mandragora turcomanica*	-	-	-	-	+	-	-	-
33	*Ophioglossum vulgatum*	-	-	-	-	+	-	-	-
34	*Ophris kopetdagensis*	-	-	-	-	+	-	-	-
35	*O. transhyrcana*	-	-	-	-	+	-	-	-
36	*Orchis fedtschenkovii*	-	-	-	-	+	-	-	-
37	*O. simia*	-	-	-	-	+	-	-	-
38	*Onobrychis nikitinii*	-	-	-	-	-	-	+	-
39	*Pistacia badfhysi*	-	-	+	-	-	-	-	-
40	*Pyrus boissieriana*	-	-	-	-	+	-	-	-
41	*P. turcomanica*	-	-	-	-	+	-	-	-
42	*Platanus orientalis*	-	-	-	-	+	-	-	-
43	*Prionotrichon haudanensis*	-	-	-	+	-	-	-	-
44	*Punica granatum*	-	-	-	-	+	-	-	-
45	*Ribes melananthum*	-	-	-	+	-	-	-	-
46	*Salsola chivensis*	-	-	-	-	-	+	-	-
47	*Sorbus graeca*	-	-	-	-	+	-	-	-

Note: Reserves: 1 – Repetek, 2 – Khazar, 3 – Badkhys, 4 – Kopetdag, 5 – Synt-Khasardag, 6 – Kaplankyr, 7 – Kugitang, 8 – Amydarya.

Table 9.6 (continued).

No	Species names	1	2	3	4	5	6	7	8
48	*Spirostegia bucharica*	-	-	-	-	-	-	+	-
49	*Stipa crassiculmis*	-	-	-	+	-	-	-	-
50	*S. kopetdagensis*	-	-	-	+	-	-	-	-
51	*S. turcomanica*	-	-	-	+	-	-	-	-
52	*Tulipa hoogiana*	-	-	-	-	+	-	-	-
53	*T. ingens*	-	-	-	-	-	-	-	+
54	*T. kuschkensis*	-	-	+	-	-	-	-	-
55	*T. lehmanniana*	-	-	+	-	-	-	-	-
56	*T. miecheliana*	-	-	-	+	-	-	-	-
57	*T. wilsoniana*	-	-	-	+	-	-	-	-
58	*Ungernia spiralis*	-	-	-	-	+	-	-	-
59	*U. victoris*	-	-	-	-	-	-	-	+
60	*Vitis silvestris*	-	-	-	-	+	-	-	-
61	*Zygophyllum bucharica*	-	-	-	-	-	-	-	+

Note: Reserves: 1 – Repetek, 2 – Khazar, 3 – Badkhys, 4 – Kopetdag, 5 – Synt-Khasardag, 6 – Kaplankyr, 7– Kugitang, 8 – Amydarya.

Table 9.7. Measures on melioration of irrigated farm lands, in ha (National Action Program.., 1996).

Measures	Velayats					Total
	Akhal	Balkan	Dashkhovuz	Lebap	Mary	
Exploitation leaching of soils	187,892	40,363	104,236	100,734	125,789	559,014
Repairing of irrigation and drainage canals	654,000	-	-	3,513	19,777	23,935
Repairing of canals and exploitation leaching of soils	39,880	257,000	108,135	78,286	125,044	351,502
Principal leaching of soils	20,621	7,619	15,177	3,009	25,978	72,404
Repairing of irrigation and drainage canals, principal leaching of soils	10,817	0,276	37,328	6,304	17,791	75,516
Construction of irrigation and drainage canals	27,495	0,917	21,518	7,051	27,454	84,435
Construction of irrigation canals and exploitation leaching of soils	142,980	2,907	92, 696	74,047	411, 608	

Protective forest belts along motor roads	1,425
Protective forest belts along railway	4,600
Sand stabilization around gas extraction stations and around other industrial objects in the desert	1,960
Stabilization of oasis sands	700.

Table 9.8. The main measures of rangeland improvement (National Action Program.., 1996)

Physical regions	Size of areas, mln. ha	Soils	Dominant plant species	Mean annual forage productivity, tons per ha	Recommended technique	Increase of productivity, %
Southwest Turkmenistan	0.7	Takyr like soil	*Salsola sp.*	0.12	Cultivation of forage plants	120
Northwest Turkmenistan	1.2	Graybrown soil	*Artemisia sp., Salsola sp.*	0.085	Water accumulation furrows	120
Transunguz Karakum	4.1	Sandy soil	*Haloxylon persicum, Carex physodes*	0.075	Rangeland reseeding	140
Central Kakakum	7.2	Sandy soil	*Haloxylon persicum, Carex physodes*	0.12	Rangeland reseeding	160
Southeast Karakum	5.0	Sandy soil	*Haloxylon persicum, Carex physodes*	0.19	Rangeland improvement	200
Loess desert (foothill country)	0.78	Light gray brown soil	*Poa bulbosa, Carex pachystilis*	0.26	Establishment of fall-winter pastures	400
Mountain regions	0.22	Typical graybrown soil	*Artemisia sp., Graminea* species	0.28	Rangeland Improvement	600

9.5. Creation of enabling environment for implementation of NAPCD

As known, implementation of NAPCD will be possible if an enabling environment is created. That includes good governance and stability, legal and administrative reforms, economic incentives and improved infrastructure.

In all countries of the region, including Turkmenistan, people are preoccupied with economic survival. The great number of refugees testifies to bad governance and instability. Legal and administrative reforms are conducted in all countries of Central Asia. Several good laws and regulations have been adopted, but the governments can't realize them because of economic disruption, corruption of the administrative bodies and the passive role of the population. According to the old communist traditions, many laws are purely declarative. Economic criteria don't play an important role, and a modern infrastructure is not created.

Decentralization and participation of the local people in realization of the NAPCD is a pledge of success. The workshop on social aspects of dryland management held in Nairobi developed recommendations which concerns governmental policy at the local level (Anonymous, 1993). One of this recommendations said: “Land reforms aimed at providing equitable access to land by those actually working it and providing them with secure, clear rights (and obligations) associated with their land tenure are essential, although not sufficient, for improving natural resource management”.

In 1992 – 1993 several legislative regulations were adopted in Turkmenistan, in particular "The Law about the Lease and Lease Regulations", "About Local Self-governmental Institutions", "Land Tenure Code", etc. New State Program "Ten Years of Prosperity" and "New Village" were also approved. The Program "Ten Years of Prosperity" was later renamed to "Ten Years of Stability" because of the lack of prosperity...

In 1991 only 41% of food requirements in Turkmenistan were covered by domestic production. More than 60% of consumer goods were imported from other republics of the former USSR or from foreign countries, including 33% of meat, 45% of milk, 70% of potatoes and 100% of sugar (Babaev & Kharin, 1994a).

Citizens of Turkmenistan can take the land on lease or in private holding, but cant sell their parcels of land. The former kolkhoz or sovkhoz were transformed to agricultural associations or to holding companies. But the government of Turkmenistan distributes to new land holders only virgin or abandoned land beyond oases. This land is not provided with roads and irrigation systems needed for normal functioning for agricultural production. So many country people prefer to work in kolkhoz or in an agricultural association. These economic structures – the former kolkhoz and new – agricultural association have slight differences. In both cases the state provides them free with irrigation water, roads and other infrastructure. The people, being slaves, don't take responsibility on land conservation and they have the possibility of being able to graze livestock illegally and cultivate vegetables on their personal land parcels. These are the main sources of income for the rural population.

Mechanisms of the market economy don't yet function in the agriculture of Turkmenistan and other Central Asian countries. For example, cotton growers working in agricultural associations or on private lots of land must sell row cotton by wholesale to state agencies by fixed prices. The state pays only 3 – 5% of the market prices. Cotton growers don't have material in centives in production of this crop.

The progress in land tenure reform in Turkmenistan is very slow. During 1992 –1994,

Figure 9.1. Stabilization of moving sand by bitumen emulsion near the motor road in the vicinity of Repetek, Turkmenistan. Photo by A. Prishchepa.

72,000 ha of virgin land were distributed among private holders. But only 1/3 of this land was put under plough. Land holders met difficulties in development of new land. Shortage of agricultural machinery, lack of irrigation water and financial constraints were the main causes of slowing down the reform. Agroprombank of Turkmenistan could give interest-bearing loans to all land holders, but high rates of interest and galloping inflation (more than 600% in 1995) hindered this activity.

Production of meat and milk in Turkmenistan in 1995 was reduced because of lack of forage and reduction of livestock numbers. According to Kasymova (1995), production of main foodstuffs in 1995 was characterized by the following figures (kg per capita per year):

	Prodution	Medical rate
Meat	25	60
Milk	113	251
Eggs (pieces)	55	158

500,000 cows in the country couldn't provide the population with milk because of low productivity of milking cows, totaling 2,350 liters of milk per head per year. The country suffered from a shortage of grain. The state program "Grain" envisaged the self-sufficiency of Turkmenistan in grain. But the program failed in 1995 because of bad management and severe drought. President Niyazov (Turkmenbashi, 1995) criticized the Ministry of Grain Products. The country wanted to purchase in Iran 200 thousand metric tons of flour and 175 thousand metric tons of grain. Later, in 1999 the problem was solved (Bablumyan, 1999).

Experiments in the spirit of "nature remaking" were conducted in Turkmenistan. Seeds of vegetables and potato were purchased in the Netherlands, the country with climatic conditions differing from that in Turkmenistan (Pavlov, 1994). The results of these experiments were deplorable. The people didn't want to consume the products because they were inedible. It was very difficult to say what dominated in this project – rank ignorance of plant breeding or evil intent?

As known, non-governmental organizations (NGOs) could play an important role in implementation of the NAPCD. There are many NGOs in Turkmenistan. Some of them were transformed from so called "public organizations" that functioned during the Soviet period. Traditionally, these organizations stayed under the full control of the government and their role in public life was very low. But there are good examples of NGO activity at a local level (Babaev & Kharin, 1994a). In the Magtymguly etrap The Union of War Veterans participated in ecological programs. N. Chakanov, the head of this organization, said that they also created a special fund of relief to help poor people.

In the Ferab etrap of Turkmenistan the local people conducted several activities on planting trees and shrubs around settlements and agricultural fields. All of the people participated in this work. A so called "monthly campaign" for planting greenery was organized. This campaign was reminiscence of campaigns conducted during the Soviet period with participation of adults and children. But in this case the people fulfilled the decree of the president of Turkmenistan on planting trees and gardens. As the correspondent noted (Grigor'eva, 1996), 2,000 seedlings of pine were planted. Many trees of poplar, maple and mulberry were also planted. Each agricultural association (former kolkhoz) had at least 50 hectares of gardens.

The role of religious organizations in solving ecological problems could also be great. The experience of Western countries proved that. N. Annaoraszov, the chief mullah of the Bakharden etrap said that several Moslem communities revived their activity forbidden during the Soviet period. 26 mosques were functioning in the etrap (Babaev & Kharin, 1994a).

Women should also participate in realization of the NAPCD. The recommendations of an international workshop on this issue said: "Mechanism of promoting women's equal participation in the decision making process at all levels should be supported by strengthening existing women's groups and organizations, or where the need is expressed to assist with their formation" (Anonymous, 1993).

These academic recommendations developed by highly educated scientists (mainly males!) couldn't solve "the problem of gender" in the region under consideration. The status of local women in villages of Central Asia was aggravated after proclaiming the independence. Slave labor on cotton plantations, bearing children, cooking and other housework didn't give any chance to participate in public activity. As known, women always suffered more during troubled times.

International cooperation

A paradoxical situation arose in Central Asia after proclaiming independence and the crash of the USSR. Economic, political and cultural ties between the former Soviet republics were broken. Each country stood apart in the limits of its territory. But the world's experience proved that only joint efforts in use of financial and material resources could solve regional problems. At last the countries of Central Asia realized the need of cooperation and created the following regional organizations:

- International Foundation for Saving the Aral Sea (IFSAS)
- Inter-governmental Council on the Problems of the Aral Sea (ICPAS),
- Executive Committee of the ICPAS.

25 statesmen (5 from each country) formed the body of the ICPAC. Two commissions were set up within the frame of the ICPAC: Commission on Water Management and Commission on Ecological Problems. IFSAS estimated the annual contribution of each country to the common fund should be about 1% of the national income. However not a single member paid what was due as yet (Anonymous, 1994a).

On the territory of Central Asia many agencies dealing with desertification are functioning, in particular: regional intergovernmental agencies, national administrations, ministries, research institutions, universities, national private companies, NGOs, UN organizations, World Bank, foreign private companies, etc. In many cases the activities of these organizations is duplicated. An army of foreign scientists, experts, consultants and advisers collect information and give promises which will never be fulfilled.

International organizations and donors developed "The Program of Needs" which was the program of the first phase of activities on rehabilitation of the Aral Sea. The Program included 20 projects with a total cost of $220 mln. Practical importance and possibility of realization of these projects will be assessed during three next years (Anonymous, 1995b). All these projects couldn't be implemented in a short time. So the local people can wait till "dooms day" to get any benefits. Maybe the next generations of people will see the results of these activities.

CCD pioneers new approaches and arrangements in international partnership, coordination and cooperation. The Regional Implementation Annex for Asia sets up a mechanism of implementation of CCD, in particular: exchange of information, cooperation and coordination of actions, promotion of scientific and technical cooperation etc. These activities includes: regional action, sub-regional (joint action) programs and NAPCD.

Central Asia being a part of Asia can be considered as a separate sub-region. This sub-region has many specific features common to all 5 countries, namely: physical conditions, historical development, culture and religion. All countries became independent in 1991 and their economy is in the stage of transition to market relations. According to the CCD, article 5, point 3 of the Regional Implementation Annex for Asia says: "Sub-regional or joint action programs may include agreed joint programs for the sustainable management of transboundary natural resources relating to desertification". This point is especially important for Central Asia because the main part of sub-region is located within the Aral Sea basin and use of shared water resources needs coordination.

Scientists and specialists have elaborated proposals on coordination of all activities to combat desertification in Central Asia (Babaev & Kharin, 1994a). These proposals could be used as a basis for the development of mechanism for implementation of CCD at sub-regional level. Sub-regional institutions may include:

- Sub-regional Council for Control of Implementation of Programs,
- Sub-regional Scientific Council,
- Sub-regional Monitoring Centers (SMC),
- National Centers of Desertification Monitoring (NCDM) in each country.

The Sub-regional Council will develop a sub-regional action program. This program should include all key elements of NAPCD, proposals on utilization of shared water resources, capacity building, education, scientific cooperation etc. Countries of Central Asia should also promote capacity building, scientific and technical cooperation on utilization of alternative sources of energy, water desalinization, development of water saving technologies in agriculture and biodiversity conservation.

The short age of water resources is the reason why the countries of Central Asia need coordination of their efforts in desertification control with joining countries: Russia, China and Afghanistan. Abundant water resources of Russia will again and again attract the attention of the Central Asian countries pour in water. Utilization of water resources of the Amudarya needs coordination with Afghanistan especially if the civil war will be stopped. As known, the city of Turkmenbashi (the former Krasnovodsk) is supplied with water delivered in tankers from Azerbaijan.

Above that, countries of Central Asia are in need of other resources, for example wood. Wood was delivered from Russia during the Soviet period.

All these problems can be solve on the basis of bilateral or multilateral agreements.

9.6. Summary and conclusion

During the Soviet period, the former republics of the Soviet Union as well as the region as a whole had a high potential to combat desertification. Scientific investigations on ecological problems were conducted in the Academies of Sciences of these republics, in Departments of Forestry and Agriculture, in universities and in state reserves. It was a harmonious (from the bureaucratic point of view !) system which stayed under the control of Moscow. A special Committee on Coordination of Scientific Works was functioning in the USSR Academy of Sciences. Training technical specialists was organized through many universities and institutions. Many leading Russian scientists participated in these activities.

But this high potential was not used in full measure because of ideological considerations. Leaders of the USSR couldn't tell the truth about the failure in realization of great projects on nature transformation. Severe censorship and corruption aggravated the situation. Ecological concerns, in the minds of ordinary people, were absent.

What happened after 1991 when the former Soviet republics became independent countries?

"The hand of Moscow" was not functioning any more. The system created over 70 years was fully destroyed. The activity of the Academies of Sciences of these republics was paralyzed because of the lack of money, low salary of scientists and corruption. Russian scientists left the territory of Central Asia *en mass*. Attempts to continue training of scientific cadres through local VAKs (High Certifying Commissions) failed. As a result the countries of Central Asia made a step back in their technical and cultural devepment.

We also want to emphasize one negative process common to all countries of Central Asia, and in general to many developing countries. The question is of the process of urbanization, the process of mass migration of the rural people to big cities. Many rural people has a dream – to move to metropolis and to solve all problems there.

For example, Ashkhabad, the capital of Turkmenistan can be mentioned here. Many multistory buildings were constructed here during the last years of the Soviet power in new city areas, so called "mikrorayon". These multistory houses looked a sorry spectacle because of absence of water, shortage of electricity and heaps of refuse around them. The governments of Turkmenistan and other countries of the region try to solve the problem of overpopulation of big cities by so called "propiska" (registration of passport), but in vain.

As known, the social ties of people in big cities are weaker than in rural country, and all kind of criminal activity is developed here. Many people has illegal status and administration can not control that. So, the new independent countries of Central Asia have the task of building up everything anew. That also means the implementation of economic reforms which will create a material basis for desertification control. NAPCDs must be developed in all countries of Central Asia. These programs will include current measures on desertification control and measures for future activities.

In conclusion, CCD is a basic international law on desertification control and mitigation of drought. This law was adopted by the world community for the first time. Countries of Central Asia don't have an alternative to sustainable development. CCD is not a panacea for all misfortunes, but it raises the hope of the peoples. A gleam of hope is seen at the end of the tunnel!

CHAPTER 10

SYNTHESIS AND CONCLUSION

Timps futurs! Vision sublime!
Les peuples son hors de l'abîme
Le désert morne est traversé,
Apres les sable, la pelouse,
Et la terre est comme une épouse;
Et l'homme est comme un fiancé!
V. Hugo – Les Châtiments

10.1. Lessons of the past

Ecological stability and environmental security of the ancient pre-Islamic civilizations in Central Asia depended mainly upon physical factors. Low productive forces and scarce population could not initiate an ecological crisis. Even, "the ruler of the world", Alexander the Great, could not destroy the productive forces of the region. The first ecological catastrophe occurred in the 6th century B.C. It was caused by climatic fluctuations and the change of river flow.

Earlier Islamic civilizations (in 8 – 11th centuries A.D.) consolidated the peoples of Central Asia. Feudal strives and wars did harm to irrigation systems at the local level but the people survived again laboring "by the sweat of their brow".

The second ecological catastrophe in Central Asia broke in the 13th century A.D. A new "ruler of the world", Cinggis Khan, subjugated many countries. The Mongol conquerors destroyed the irrigation systems and raised to the ground many flourishing oases in Central Asia. Some of them were never restored again. The people of Central Asia spent two centuries of heavy labor for rehabilitation of the national economy and culture.

The next three centuries (the 16 – 18th) in the history of Central Asia were marked by fluctuations from prosperity to decay. By the end of the 18th century the political situation and economy were stabilized and the Islamic civilization flourished.

Bukhara and Samarkand were centers of Islamic culture. During this period the ecological situation always stayed under the control of the people.

The political stability in the region didn't changed in the 19th century. In spite of the contacts of the local people with Europeans, the Islamic civilization conserved its originality. Traditional technology in agriculture did not change, even after the Russian conquest. Economic prosperity and ecological stability were characteristic features during the rule of "Ak Padishakh" in Russian Turkestan.

A new era in the history of Central Asia, the era of "Socialism" began in 1917. The Bol'shevik revolution brought new misfortunes to the people of Central Asia. Millions of people were killed or died from hunger in the Civil War, by collectivization and in clashes with basmach bands. But in spite of all that, great progress was achieved in material sphere and education. General literacy replaced medieval ingnorance. But the end of this civilization was marked by the third ecological catastrophe in Central Asia.

Construction of the Karakum Canal and use of the river flow of the Amudarya and the Syrdarya for irrigation caused a drop in the Aral Sea level. A new anthropogenic desert was formed on the dried up bottom of the Sea. Salts blown off from this area were transported thousands kilometers. Some analysts believe that it was the beginning of a new global ecological crisis.

In 1991 new independent countries appeared on the geographical map of Central Asia. These countries inherited the problem of desertification from the former Soviet Union. Accumulated Ecological Problems become serious bottlenecks for sustainable development of these countries.

In spite of all shortcomings of the Soviet political system, several new techniques were developed for regeneration of forests, rangeland improvement and soil melioration during the Soviet period. These techniques being developed by Russian and local experts were considered during the Soviet period as great achievements of science and technology. If their ideo- logical motivation will be rejected, the new administration can use these techniques in solving ecological problems. The continuity of ecological policy will give a positive effect.

10.2. Up-to-date situation

At present, peoples of the world have several global problems: geopolitical, economic, demographic, ecological, etc. Many analysts believe that these problems can't be solved by national governments. Only joint efforts of he world community can solve these problems. Desertification being a global process belongs to these problems. The trend of these events is known as "globalization". From this point of view, the United Nations Convention to Combat Desertification (CCD) is a good example of solving a global problem. As to Central Asia, we have come across with other problem – the problem of souverenization. The crossing of these two problems makes additional difficulties in desertification control.

The drop of the Aral Sea level has changed the ecological situation in Central Asia. Desertidication embraced all land use types. 7 – 10% of desert rangelands lose their productivity annually. Waterlogging desert rangelands in the oasis bordering zone was registered in the area of 34.4 thousand km^2. The situation is especially severe in Kazakhstan where forage production was exhausted by the 1960s.

The area of saxaul forests has reduced by 10 – 12% annually. About half of the areas occupied by mountain forests have low productivity and low crown closure. They belong to the category of understocked forests. These forests can't fulfil their protective functions. The size of the area of tugai forests was considerably reduced during the last decades.

The deficiency of timber in the region is a serious bottleneck in sustainable development. During the Soviet period, this deficiency was covered on the account of timber delivered from Russia. The goverments of Central Asian countries must avoid the temptation of solving this problem by exploitation of limited forest resources.

Severe wind erosion affected vast areas, especially in Kazakhstan where 20.5 mln. ha of plough land and 25 mln. ha of rangelands were affected.

Cotton plantations in Central Asia occupy about 60% of irrigated land. From them, 13 – 20% are severely affected by salinization. Above that, 2 – 3 % of irrigated soils lose their productivity annually.

Destruction of the vegetative cover, reclamation of new lands and human pressure on environment affected the reduction of biological diversity of plants and animals. Game – shooting and poaching were the reasons why the number of animals, especially hoofed animals and snakes was reduced.

By rough estimate, the direct economic damage of desertification for all countries of Central Asia totals about $ 1.7 billion.

Desertification, being a destructive process, hits the physical but also the social environment. Populations of the areas affected by desertification suffer from bad quality of water. The worsening ecological conditions cause increase of respiratory and infectious diseases. High infant mortality and the reduction of the length of life are also the consequences of desertification.

The problem of refugees is also very actual for Central Asia. Some 4 – 5 mln. of people left the territory of these countries since 1999 because of political instability, poverty and desertification. According to forecasts the flow of people leaving Central Asia will continue during the next years. Many of the refugees are technical specialists, doctors or workers of high qualification. Their migration undermines the efforts of local administration on reconstruction of national economy and desertification control.

10.3. The requirements of future activities in desertification control

According to forecasts of scientists, aridity in Central Asia will increase during the next 25 – 30 years. The flow of all rivers to the Aral Sea will total only 22% of that in 1987. This amount of water will cover only 20 – 25% of water demand. So, we may expect that scarcity of water resources will become a major ecological and political problem in the region.

During the last 100 years the population of Central Asia increased 10 times. In 1994 it totaled 53.4 mln. About 110 mln. people will live in Central Asia by 2020, if the existing rate of population growth will be not changed. Political leaders of Central Asia following the traditions of Islam don't want to take measures on birth control. So, shortage of food will also be a sharp political problem.

The most probable path that Central Asian countries will follow is the line of least resistance. That means that these countries will join Islamic world. Poverty and low level of technology will equalize them with adjacent Moslem countries.

What to do? Will the epoch of Noosphere, the epoch of full harmony between man and environment come in the nearest future in Central Asia? That is possible, but hardly probable... CCD gives the gleam of hope in implementation of NAPCDs. Several legislative regulations – laws on nature conservation, land tenure codes and other acts – have been adopted in all countries of Central Asia. They can provide in future the implementation of NAPCDs. Slightly degraded lands mainly need conservation of natural resources. Moderately degraded lands need conservation and partial rehabilitation of ecosystems. Severely degraded lands can be restored only by special ameliorative measures. New approaches and new technologies will play an important role in these activities.

These realistic tasks in rehabilitation of degraded lands could be achieved,

- if political leaders and decision makers will in their activities follow the principles of sustainable development,
- if "top-down" approach will be replaced by "bottom-up" approach,
- if decision makers will find some money in public coffers for implementation of priority projects,
- if foreign donors will give some money and if corrupted administration will not plunder this money,
- if the people will overcome their traditional conservatism and enmity to innovation,
- if the people will get over the threshold of poverty and will pay more attention to ecological problems,
- if the people will have more economic and social incentives by conducting measures on desertification control,
- if ecological education and awareness of desertification will embrace all classes of society,
- if...

The author is a little optimistic. Man being a wise creature of God can survive only in common with nature. There is no alternative to anti-desertification activities. Soon or later desertification in Central Asia as well in the whole world will be put under the control of people. And addressing the readers of this book the author wants to conclude using the lines of N.A. Nekrasov:

That's a pity,
But neither me nor thou
Will live so long
To see these blessed days!..

(Translated from Russian by N. Kharin)

Black saxaul in the Repetek reserve, Turkmenistan

GLOSSARY OF LOCAL AND SPECIAL TERMS

Active wind, erosion wind, sand wind the wind which produces the transportation of wind particles, and which can cause **deflation** or **wind erosion** (See). Active wind is the wind with the velocity of 5 m/sec and more measured at the height of 10 m above the ground (the height of installing the weathercock)

Adat code of tribal traditions and rites, "unwritten code" of Moslems

Adyr hill country in Turkmenistan

Afforestation of sands the artificial establishment of forest crops in sandy areas

Afganets a local southeast wind in the upper Amudarya region (Turkmenistan)

Agricultural drought shortage of moisture in the soil hindering the growth and ripening agricultural crops

Agricultural spring drought deficiency of moisture in the soil during the growth of cereals

Agricultural summer drought deficiency of moisture in the soil during the growth of later crops

Agricultural autumn drought deficiency of moisture in the soil during the sowing and growth of winter crops

Agro-forestry practicing agriculture and forestry together. Establishment of shelter belts and forest plantations is an important measure which ameliorate the microclimate and increases the productivity of agricultural crops

Air drought deficiency of moisture in the air

Ak gui well dug in a sandy area (Karakum Desert)

Aklang sand tract surrounding wells, devoid of vegetation (Karakum Desert)

Aksakal (lit. "white beard") old man

Aladzha color thread tided around the wrist of children on the purpose to protect them from evil spirits

Alternative livelihoods new livelihoods to be compared with traditional livelihoods earned by local people for a long period of time in the locality under consideration

Alternative energy system new energy systems, for example solar, wind, biogas, etc. Which don't contaminate the environment and which could replace traditional energy systems

Anasha drug produced from hemp

Anchored dune sand dune covered with vegetation which has arrested its movement

Aral climate dry continental climate with severe winter, hot summer and spring precipitation (Central Asia)

Arba primitive cart, consisting of a small platform and two big wheels

Archa trees and shrubs of *Juniperus* species

Arid climate 1. A climatic type which is characterized by insufficient moisture to support appreciable plant life. 2. Generally, extremely dry climate

Aridity 1. Conditions in which evaporation always exceeds precipitation. 2. Characteristic of a climate relating to insufficiency or inadequacy of precipitation to maintain vegetation

Aridity index ration of available precipitation to potential evapotranspiration. According to this criteria the climates are classified as:

Hyperarid (Extra-arid)	< 0.05
Arid	0.05 – 0.20
Semi-arid	0.20 – 0.50
Dry sub-humid	0.50 – 0.65
Moist sub-hunid and humid	> 0.65

Arid soil common term for soils formed under the conditions of arid and semi-arid climates where evapotranspiration exceeds precipitation

Arid zone 1. Zone in which precipitation is so deficient that irrigation must be practiced to support cultivation. 2. Zone in which evaporation always exceeds precipitation
Aryk irrigation canal
Aryk aksakal official responsible for distribution of water from the main irrigation canal to small aryk (See) in Russian Turkestan
Aul village, settlement

Badland land with rough topography having steep barren slopes and narrow gullies, most common in arid and semi-arid regions
Bai rich man, the same as kulak (See)
Baibak (lit. " sleepy head") marmot
Bakshish bribe, gift giving to officials
Barkhan crescent sand dune. Windward slope of barkhan is gentle and lee slope (slip face) is at the angle of repose of dry sand
Basin irrigation a type of irrigation whereby irrigation water is directed into "basins", surrounded by earth banks
Basmach member of guerilla bands which fought against the Soviet administration
Beg high official, the head of province in the Bukhara Khanate
Blowout trough or hollow formed in sandy area by the action of wind
Blowout dune sand accumulation derived from blowout troughs or depressions, especially where the accumulation is of large size and rises to considerable height above the source area
Bogara agriculture in arid regions conducted without irrigation
Bogaz (lit. " throat") passage, mountain pass, strait
Brackish slightly salty (applied to water which is neither fresh nor salty, but in between)
Brush matting a matting of branches placed on badly eroded land to conserve moisture and to reduce erosion. In Central Asia this method is used for creation of mechanical barriers in chess board pattern for stablization of moving sands

Capacity building to combat desertification institution building, training and development of relevant local and national capacities (Convention to Combat Desertification)
Central Asia 1. In Russian geographical literature, physical region which includes Mongolia and the northern part of China. 2. Syn. Middle Asia (See) . 3. Geopolitical region which includes five newly independent countries in the limits of their national boundaries (Kazakhstan, Uzbe kistan, Kirghizstan, Tajikistan and Turkmenistan). The term was put in use after 1991.
Chaban herdsman, shepherd
Chagyl slightly undulated sandy steppe (Kazakhstan)
Chaikhana tea house
Chal drink made of camel milk
Chemical sand stablization stabilizing sand dunes with oil or other chemicals that form a crust on the sand surface. Often this measure is combined with establishment of forest crops
Chigir' a kind of wheel to lift water from canal to irrigated fields
Chink any abrupt, dry except in the rainy season
Chirly well constructed on takyr (See). The well is a part of runoff-inducement system used in the Karakum Desert Chul steppe or land devoid of water
Collectivization formation of **kolkhoz** (See)
Combating desertification activities which are part of integrated development of land in arid, semi-arid and dry sub-humid areas for sustainable development which are aimed at:

1. Prevention and/or reduction of land degradation
2. Rehabilitation of partly degraded land
3. Reclamation of desertified land (Convention to Combat Desertification)

Commonwealth of Independent States (CIS) community of the former Soviet republics which became independent after the desintegration of the USSR. The Baltic states are not members of CIS
Continuous grazing a specific area throughout a grazing season (for example a season of year), that provides the best forage or is suitable because of presence of water, etc.
Contour stripcropping cultivation of agricultural crops along contour lines
Coppice mound (Coppice mound dune) accumulation of sand and dust around desert plants

Dacha country house
Dag a mountain or hill (Turkmenistan)
Dam – dam a spring giving water drop by drop
Damla well constructed in sand underlain by porous rock
Dar'y a big river (Central Asia)
Dashkak a water cistern made of bricks (Karakum desert)
Deflation removal of sediments by the wind
Deflation armor See Desert pavement
Deflation basin a hollow formed by removal of sand and dust by aeolian action
Deforestation removal of forest crop without intention of reforesting.
Degradation of vegetation diminishing the biological productivity, change of plant composition and loss of economic potential of natural vegetation
Deigish erosion of river banks
Dekhanin or deikhanin farmer (Plural **dekkhane)**
Depe hill
Desert climate a climate type which is characterized by insufficient moisture to support appreciable plant life, that is a climate of extreme aridity
Desertification land degradation in arid, semi-arid and dry sub-humid areas resulting from various factors, including climate variations and human activities (Convention to Combat Desertification)
Desertification criteria quantitative or qualitative indices which characterize the process of dry land degradation, e. g. decrease of pasture productivity, kg/ha, removal of top soil, cm, etc.
Desertification indicators physical or biological processes in arid ecosystems which indicate the degradation of dry lands, e. g. the loss of biological productivity, soil degradation etc.
Desertification map thematic map which shows by special symbols the status or other characteristics of desertification for the territory to be mapped
Desertification monitoring the system of assessment and control of desertification based on ground observations and application of remote sensing technology
Desert patinat thin layer of manganese and iron, of dark brown color, covering the stony surface of the desert
Desert pavement coarse, angular to sub-rounded rock fragments on undissected portion of alluvial fans, terraces or valley flats
Desert polish smooth and polished surface of rocks formed by the action of wind and blown sands
Desert strip farming cultivation of agricultural crops on terraces or strips of terrain where runoff is accumulated
Desert rangeland See **rangeland**
Desert strip farming cultivation of agricultural crops on terraces or in strips of ground where runoff is accumulated
Desert woodland in Central Asia, communities of **saxaul** (See) and other desert shrubs
Desiccation a natural process by which a region suffers complete loss of water owing to decrease in rain, increase of evaporation or change in other climatic factors
Dessiatina old Russian measure of land, 1.09 hectare
Drought threshold low content of soil moisture corresponding to the wilting point when agricultural crops wither up
Drylands lands with low precipitation rate where transpiration exceeds precipitation
Duab of Turkestan (an obsolete term) the territory between two big rivers of Central Asia: the Amudarya and the Syrdarya
Dune a hill or ridge of drifted sand piled up by the wind. In the wide meaning of the word, any accumulation of sand. The term **dune** in the Soviet geographical literature was mainly applied to non-desert sand formations
Dune plants species of plants confining to dune sites. Some of them can be used for sand stabilization.
Dunganin Chinese adopted Islam (Central Asia)
Dust bowl in semi-arid regions, areas from which surface soil is being or has been removed by the wind.
Dust devil, dust whirl swirling, vertical updrift of air developed by local heating of the air avove the flat desert floor
Dust storm the strong wind carrying clouds of dust or silt and larger than **dust devil** (See)
Duval fence made of clay
Dzhut loss of cattle caused by shortage of forage

Endangered species plant species whose rehabilitation is possible only with application of special measures

Environmental refugees people who have left their homeland because of environmental disasters, like drought, desertification, flood etc.

Eolomotion all forms of surface motion of eolian material

Estivation a process of passing a hot season in an active state

Etrap administrative district in Turkmenistan, a part of velayat (See)

Exotic river a river which appears in a humid region, flows across a desert and reaches the sea or a more humid land.

Exsiccation drying up an area without reducing the rainfall (for example draining a marsh)

Falling dune sand accumulation on a slope

Furrow dams small embankments between furrows which conduct irrigation water

Field stripcropping cultivation of crops in bands located along the slopes

Fixation of moving sands all measures on stabilization of moving sands by plants, chemicals or placing mechanical barriers

Forest land land on which the native vegetation is forest. In the former USSR, it included **stocked forest land** (See) and **understocked forest land** (See) and forest cuttings

Forest planting in silviculture, planting trees and shrubs. In arid regions various methods of planting are used, in particular: ball planting, basket planting, group planting, improvement planting, mound planting, pit planting, pot planting, spot planting trench planting, etc.

Gindan prison

Glasnost' campaign of publicity proclaimed by M. Gorbachev

Goladzhy a well constructed in big elongated depressions. It usually contains salt water (Karakum Desert)

Gosplan State Commission for Planning

Grazible woodland in Central Asia, land covered with **saxaul** (See) communities and used for used for grazing without significantly impairing forest values

Hydrological drought shortage of surface and underground water resources caused by prolonged **meteorological drought** (See)

Ilak local name for desert plant *Carex phy-sodes* (Turkmenistan)

Improvement of desertified lands measures on partial amelioration of desertified lands by soil conservation, afforestation, pasture improvement, etc. aiming to increase their potential in the process of current activity of people

Increaser plant species plant species number of which is increased after **overgrazing** (See)

Integrated development coordinated economic and social development embracing all branches of the national eco-nomy on the basis of one concept

Invader plant species plants absent in local flora and invading pastures (rangelands) after overgrazing

Irrigation artificial application of water to lands for agricultural purposes

Jailow high mountain rangeland in Central Asia

Kak or **khak** pit for collection and conservation of water in the desert

Kalym bride money

Karakharsang dense layer of moss (usually of *Tortula* genus) on the desert surface

Karakul astrakhan

Karakum (lit. "black sand") sands grown by vegetation, also the name of desert in Central Asia

Kaurma mutton roasted in fat

Ketmen' a special hoe used in Central Asia since ancient time

Khalat long outdoor clothes of men and women

Kharman storing place of cotton

Khauz small pond for keeping water

Kibitka primitive dwelling

Kishmish currant

Kolkhoz collective farm

Koshma thick felt

Kozak or **Cossak** a member of the military estate in Russian Empire

Koshara shed for protection of livestock

Koumiss fermented mare's milk
Kulak (lit. in Russian "feast") rich man exploiting poor people
Kum sand, tract of sandy desert (Central Asia)
Kurdyuk fatty tail of sheep
Kyariz (or **karez**) underground tunnel dug in order to intercept the ground water and conduct it to irrigated fields
Kyzylkum (lit. "red sands"), also the name of the Kyzylkum Desert

Land the terrestrial bio-productive system that comprises soil, vegetation, other biota, and the ecological and hydrological processes that operate within the system (Convention to Combat Desertification)
Land reclamation human activity on land utilization, conservation and improvement for the benefit of man
Landscape method of photo interpretation a method of thematic interpretation of aerial and space photos based on preliminary identification of landscape units
Lepeshka flat bred
Local community community of people usually living in one village united by common historical traditions, culture, a common land use system and forming, as a rule, one administrative unit

Management of desertified lands system of measures on proper utilization of desertified lands aiming to stop their further degradation
Mass exodus from rural areas affected by desertification mass movement of rural people from the areas severely affected by desertification
Mazar tomb
Melkosopochnik land relief consisting of highly dissected hills, shallow valleys and ravines (northern Kazakhstan)
Merab in Russian Turkestan, a person responsible for distribution of irrigation water between the fields
Meteorological drought long drought caused by absence of precipitation
Microcatchment farming cultivation of agricultural crops in little patches where rain water is accumulated
Middle Asia physical region which includes Uzbekistan, Kirghizstan, Tajikistan, Turkmenistan and the southern part of Kazakhstan. The northern boundary of the region coincides with the boundary of the desert zone. The term was mainly used by Soviet geographers.
Migration mass movement of people from their domicile under pressure of poverty, political instability or ecological disaster
Migratory dune any dune which moves or is displaced by the activity of wind
Mitigating the effect of drought activities related to the prevention of drought and intended to reduce the vulnerability of society and natural systems to drought as it relates to combating desertification (Convention to Combat Desertification)
Mulla Moslem priest

National Action Program to Combat Desertification (NAPCD) according to Convention to Combat Desertification all affected countries commit themselves to develop a long term strategy to combat desertification. NAPCD is the principal document which contains these obligations
Natural pasture See rangeland
Noosphere sphere of high intellect, a stratum on the earth in which the human society will develop in full harmony with enironment
Nova a special device for lifting water from irrigation canal to the field

Oasis sands sand dunes confined to the transitional zone between the desert and oases. In Turkmenistan this zone is 15 km in width
Obkom Regional Committee of the Communist Party of the USSR
Oitak dry agriculture in shallow depressions (Turkmenistan)
Omach primitive wooden plough with an iron ploughshare
Open herding free ranging of herds while grazing, in the deserts of Central Asia it is the only method of grazing
Otara flock
Overgrazing continued overuse of grazing lands resulting in loss of pasture productivity and change of the composition of plant species
Overstocking temporary overuse of grazing land that can lead to **overgrazing** (See)
Oyuk in Turkmenistan, a road sign, usually a heap

of stones or a cone made of earth and branches

Papakha tall hat, usually of sheep skin

Parandzha special veil to cover the face of Moslem woman

Pasture 1. A grazing area enclosed and separated from other areas by fence. In Central Asia such areas are practically absent, except some small experimental sites, so the term **rangeland** (See) is preferable. 2. Forage plants used as food by grazing animals.

Pasture protection shelter belt protective belt of trees and shrubs planted between different units of grazing lands. In Central Asia, local species and newly introduces species were used

Peresortitsa falsification of cotton growth grade by storekeepers in storing places

Perestroika restructuring the economy and political system of the USSR proclaimed by M. Gorbachev

Phenolody periodic changes of biological phenomena of plants

Phenological phases seasonal stages of biological development of plants, like germination, leafing out, leaf coloration, leaf fall etc.

Phytomelioration a widely used term in Central Asia which includes all types of application of plants for stabilization of moving sands and other types of melioration works

Politbyuro Political Bureau of the Central Committee of the USSR Communist Party

Pood an old Russian measure of weight, 16.38 kilogram

Preferred species plant species that are preferred by animal grazing

Prevention of desertification measures to prevent the prolonged deterioration of biological and economic potential of lands prone to desertification

Prilavki foothills joining the Tian–Shan mountains

Pripiska (Russian) falsification of reports and statistical information

Propiska (Russian) permit for residence with the obligatory registration of passport in police

Raikom District Committee of the USSR Communist Party

Rain–fed agriculture cultivation of agricultural crops during a rainy season

Rangeland in Central Asia, land on which desert plants (shrubs, semishrubs, grasses and forbs) are used for grazing and browsing. The term should be not mixed up with **pasture** (See)

Rangeland classification in Central Asia the following main types of rangelands are identified, depending on plant composition, soil, forage quality and man's influence:

1. Rangeland types - rangelands with similar site conditions, the same plant association and similar forage quality.
2. Anthropogenic subtype of rangeland - a modification of a rangeland type differing on man's influence.
3. Group of rangeland types – combination of pasture types with similar dominants and the same edificator.
4. Class of rangelands – combination of group of rangeland types within the same landscape, e. g. rangelands of sandy desert, rangelands of gravel desert, rangelands of river valleys etc.

Range readiness conditions of rangelands providing a limited grazing without any damage to vegetation and soil

Raskulachivanie (Russian) eradication of kulak (See) as a social class proclaimed by J. Stalin

Rehabilitation of desertified lands partial or full restoration of biological and economic potential of lands prone affected by desertification

Rill wash erosional action of ephemeral concentration of surface flows of water

Rishta (lit. "thread") name of disease caused by tropical worm

Risk of desertification inherent risk of desertification, the sum of physical and ecological factors which determine the ecosystem stability

Runoff harvesting, runoff inducement collection of runoff water

Rural beatification conservation and melioration of esthetic features of rural landscape

Russian speaking people national minority consisting of the Russians and other alien people in the countries of CIS (See **Commonwealth of Independent States**)

Sagene old Russian measure of length, 2.13 meter

Sagene, cubic old Russian measure of volume, 9.66 m^3

Saman 1. Building material, bricks made of clay and straw and dried in the sun (Central Asia). 2. Dried up ephemeral vegetation (Turkmenistan)

Sand binders, sand binder plants plants which stabilize the sand surface by the activity of their roots, accumulation of organic matter and used in sand stabilization works

Sardoba or **sardaba** brick cistern covered with a dome-like roof

Sargan cheese produced from sheep milk (Turkmenistan)

Saxaul common name for two species of desert shrubs: 1. **Black saxaul** for *Haloxylon aphyllum*. 2. **White saxaul** for *Haloxylon persicum*

Sel' or **sil'** catastrophic mud avalanche in the mountains of Central Asia

Semi-arid zone 1. Zone where in some years the precipitation is insufficient to maintain cultivation. 2. Zone where evaporation frequently exceeds precipitation

Serios (severe) drought long drought lasting more than one year (or one season) and causing catastrophic disaster in agriculture

Shaitan arba (lit. "devil's vehicle") nickname of locomotive given by natives of Central Asia who saw it for the first time

Shariat code of religious, ethic and lawful directions of Islam

Shiblyak thickets of xerophylous shrubs and trees in the mountains of Central Asia

Soil erosion removal of soil by forces of nature more rapidly than various soil forming processes can replace it, particularly as a result of man's ill-judged activities such as clearance of vegetation and up-and-down slop cultivation

Solonchak white alkali soil

Solonetc black alkali soil

Sovkhoz state farm

Speed of desertification rate of qualitative or quantitative change of desertification status per given period of time, usually 10 - 15 years

Stability threshold ecosystem status by which it could lose the ability for natural regeneration

Stanitsa settlement of **kozaks** (See)

Steppe (Russian) a wide treeless plain of grassland

Stocked forest land in the USSR (and in Russia) forest land with volume density >0.2, in the deserts –woodland with crown closure >0.2

Subbotnik in the former USSR, labor freely given to the state on days off or overtime

Sukhovei dry wind (Russian)

Sustainable rural development economic and social development of rural country based on the concept of the continuity and constancy of utilization or renewal of resources without their exhaustion

Syrt elongated elevations, slightly rolling topography

Takyr shallow clay-filled depression in the deserts of Central Asia

Talent ancient monetary and weight unit, totaling 16 to 25 kilogramme

Technogenic desertification degradation of dry lands caused by consrtuction of road, pipelines, mining and other technical activities of man

Technogenic sands shifting sands originated in the places of human activities, like construction of canals, roads etc.

Technology for combating desertification the sum of methodologies and technical means used for realization of measures to combat desertification

Teriak drug produced from poppy

Toi bridal banquet or any festival

Traditional technology for combating desertification the sum of traditional measures proposed by local people to combat desertification

Tugai thickets of trees, shrubs and grasses in river valleys of Central Asia

Turkmenian climate dry continental climate with mild and a little rainy winter, dry and hot summer. Extension: region to the north of the Iranian mountains

Tyrlo place around desert wells avoid if vegetation, where livestock is concentrated for watering and rest

Ukase (Russian) decree, edict

Undergrazing light grazing which can't cause the degradation of desert rangelands in Central Asia. In this case, domestic animals don't loose the soil surface, and forage plants don't regenerate. As a result **kara-kharsang** (See) is formed

Understocked forest land in the USSR (and in Russia) forest land with volume density <0.2, in the deserts – woodland with crown closure <0.2

Uyezd administrative district in Russian Empire

Uraza fasting time of Moslems

Uryuk apricot

Uzboi a relic dry river valley, also the name of the ancient Amudarya channel

VAK in the former USSR (and in Russia) the State Certifying Commission

Velayat an administrative region in Turk-menistan

Vershok old Russian measure of length, 4.45 cm

Versta old Russian measure of length, 1.07 km

Water conservation the control, protection, management and rational use of water resources for the maximum benefit for man

Water erosion removal of soil and rock material by water. Convention to Combat Desertification describes water erosion as one of the processes of land degradation

Windbreak 1. A shelterbelt or other obstacle maintained against the wind. 2. Wind breakage , for example the breaking of trees or branches by wind

Wind erosion removal of soil and rock material by wind. Convention to Combat Desertification describes "soil erosion caused by wind/ or water" as land degradation arising from human activities

Wind shadow that portion of slope, cliff or escarpment which is protected from the direct action of the wind blowing over it

Windthrow uprooted by wind

Woodland a land occupied by trees and shrubs (See **desert woodland**)

Woodland range woodlands also used as range (See **desert rangeland)**

Yearlong grazing continuous grazing for a calendar year (usually practiced everywhere in Central Asia)

Yurta felt tent, mobile dwelling

Zakaznik closed area for conservation of rare species of plants and animals

Zastoi (Russian) period of stagnation in the USSR, usually identified with the rule of L. Brezhnev

REFERENCES

Abdraimov, C., 1990. Theoretical and practical aspects of exploitation and improvement of desert rangelands in Kazakhstan. Summary of D Thesis. Desert Research Institute, Ashkhabad, 45 p. (Russian)

Agabaev, A.,1996. Looking at the face of life. Neitral'nyi Turkmenistan, 5 April: 1. (Russian)

Aizin, B. M., 1986. Distribution, number and prospects of dzheiran protection in Kirghizia. In: Protection and prospects of rehabilitation of dzheiran in the USSR, p. 39 - 91. Gosagroprom Publ., Moscow. (Russian)

Akhovetsky, V. I. & Bogdanov, I. Ya., 1988. Kara-Bogaz-Gol yesterday, today and tomorrow. Ashkhabad, 340 p. (Russian)

An, P. A. & Konovalova, N. S., 1981. On methodology of agrometeorological assessment and forecast of productivity of main types of rangelands in Uzbekistan. Proc. Central Asian Institute of Agrometeorology, 83 (164): 38 - 44. Gidrometeoizdat Publ., Moscow. (Russian)

Anonymous,1891. Bukhara. Encyclopaedia Brokhauz & Efront Publishers, Vol. X, p. 97 - 110. Saint Petersburg. (Russian)

Anonymous, 1902a. Khiva Khanate, p. 184 -194. Encyclopaedia. Brokhauz & Efront Publ., Vol XXXVII, Saint Petersburg. (Russian)

Anonymous, 1902b. Turkestan, p. 174 - 203. Encyclopaedia. Brokhauz & Efront Publ., Vol. XXIV. Sant Petersburg. (Russian)

Anonymous, 1985a. Let's intensify the campaign against falsification. Izvestiya, 22 April: 2. (Russian)

Anonymous, 1985b. Karakul labyrinth. *Izvestiya*, 5 June: 1. (Russian)

Anonymous, 1988a. In the Procurator's Office of Turkmen SSR. *Turkmenskaya Iskra*, April:1. (Russian)

Anonymous, 1988b. Advocates take the floor to speak. *Izvestiya*, 23 December: 1. (Russian)

Anonymous, 1989. In the Procurator's Office of Turkmen SSR. *Turkmenskaya Iskra*, 30 April: 1. (Russian)

Anonymous, 1990. In the Procurator's Office of the USSR. *Izvestiya*, 9 February: 1. (Russian)

Anonymous, 1992. Nuclear explosions resounded in 27 places of Kazakhstan. *Izvestiya*, 29 October: 3. (Russian)

Anonymous, 1993. Listening to the people: social aspects of dryl and management.

Proc. Int. Symposium on Sustainable Development, 18 December 1993. Nairobi, Kenya, 12 p.

Anonymous, 1994a. Aral Sea: an emergency programme approved. *Turkmenskaya Iskra*, 25 June : 2. (Russian)

Anonymous, 1994b. No alternative to skilful agriculture. *Turkmenskaya Iskra*, 11 July: 1. (Russian)

Anonymous, 1995a. Activity of commission on reorganization of collective farms was ineffective. *Turkmenskaya Iskra*, 3 September: 1. (Russian)

Anonymous, 1995b. Global project is realized in practice. *Vestnik Arala*, 1 August: 1. (Russian)

Anonymous, 1995c. Joining up Central Asia for saving the Aral Sea. *Vestnik Arala*, 1 August: 1. (Russian)

Anonymous, 1996. Threat of religious clashes in Kirghizia. *Segodnya*, 1 July: 4. (Russian)

Anonymous, 1997a. Capital punishment for narcotics trade. *Izvestiya*, 23 January:1. (Russian)

Anonymous, 1997b. A deluge threatens the Pamir. *Novye Izvestiya*, 30 December: 1,2. (Russian)

Anonymous, 1998. 91 persons were killed during flood in Uzbekistan. *Izvestiya*, 15 July : 3. (Russian)

Anonymous, 1999a. Askar Akaev: The economy of Kirghizia is destroyed by activity of criminals. *Izvestiya*, 30 July: 4. (Russian)

Anonymous, 1999b. Ten million dollars to control locust. *Izvestiya*, 3 November: 4. (Russian)

Anonimous, 2000. Wages in Kirghizstan are de-

creased. *Izvestiya*, 20 January: 4. (Russian)
Antonov, V., 1990. What are the problems of Aral? *Zvezda Vostoka*,12 August: 3–9. (Russian)
Arabaev, A., 1994. Looking in the face of life. *Neitral'ny Turkmenistan*, 5 April: 1. (Russian)
Aranbaev, M. P., 1969. Sierozem and light chestnut soils of the central Kopetdag. Ylym Publ., Ashkhabad, 175 p. (Russian)
Ardaev, V. N., 1993. N. Nazarbaev proposes to introduce again the five-year plan system. *Izvestiya*, 16 January: 4. (Russian)
Ardaev, V. N., 1997. Kazakhstan could be transformed to "Central Asian snow leopard". *Izvestiya*, 11 October : 2. (Russian)
Atamamedov, N. V. & Moshev, M., 1984. Combat for reconstruction of the national Economy, p.118 – 121. TSSR. Encyclopeadia, Ashkhabad. (Russian)
Azizova, G. & Khasanov, U., 1999. State capitalism of Islam Karimov. *Izvestiya*, 24 December: 4. (Rissian)
Babaev, A. G. & Kharin, N. G.,1994a. International Convention to Combat Desertification. Turkmenistan preparatory activities. Desert Research Institute, Ashkhabad, 39 p.
Babaev, A. G. & Kharin, N. G.,1994b. International Convention to Combat Desertification. Sub-regional preparatory activities in Central Asia. Desert Research Institute, Ashkhabad, 38 p.
Babaev, A. G. & Kharin, N. G., 1994c. Report on state of environment in Central Asian republics. Desert Research Institute, Ashkhabad, 75 p.
Babaev, A. G., Kharin N. G. & Orlovsky, N. S.,1988. Complex desert development and desertification control in the USSR, In: Proc. Intern. Res. And Development Conf., Tucson, USA, 20-25 October 1985, London. p. 825 – 839.
Babaev, A. G., Kharin, N. G. & Orlovsky, N. S., 1993. Assessment and mapping of desertification processes. A methodological guide. Ylym Publ., Ashkhabad, 63 p.
Babaev, A. G. & Laptukhov, A., 1992. Phenomenological study of correlation between the Caspian Sea level with the solar activity. *Problemy Osvoyeniya Pustyn'* 5: 3-12. (Russian)
Babakhanov, U.,1992. Tajikistan. A new leader is acting on the proverb: honey is sweet, but the bee stings. *Komsomol'skaya Pravda*, 16 December: 4. (Russian)
Bablumyan, S., 1999. The share of Russian and Kazakhstan in production of grain in CIS is falling, and the share of Ukraine is increasing. *Izvestiya*, 17 August: 4. (Russian)
Bairamov, D. B.,1988. A project of development of the production forces of the Turkmen SSR until 2005. Part I, Ashkhabad, 203 p. (Russian)
Baitin, A. A., Baranov, N. I. & Gernits, O. O. et al., 1950. Principles of forest management. Goslesbumizdat Publ., Moscow, 463 p. (Russian).
Baitulin, I. & Bekturova, G., 1997. National strategy to combat desertification in the republic of Kazakhstan. *Desertification Control Bulletin* 30: 19 - 27.
Bartol'd, V. V., 1900. Turkestan during the epoch of Mongolian conquest. Saint Petersburg, 178p. (Russian)
Bartol'd, V. V., 1914. About history of irrigation in Turkestan. Saint Petersburg,174 p. (Russian)
Batygin, A., 1993. The dinner from "The Red Data Book". *Pravda*, 14 January: 4. (Russian)
Batyrov, A., 1990. The types of economic development and peopling the territory of Turkmenian deserts, p.122 – 129. In: The scientific basis of economic and social development in arid regions, Ashkhabad. (Russian)
Batyrov, A., 1992. Not repeat the mistakes of the past. *Turkmenskaya Iskra*, 20 November: 3. (Russian)
Batyrov, Sh., 1950. The thanks of Turkmen people. *Pravda*, 14 September: 1. (Russian)
Bereznev, E. N., 1995. Improvement of desert rangelands in the Aral Sea bordering areas. *Problemy Osvoeniya Pustyn'* 1: 43-47. (Russian)
Bie, S. W. & Imevbore, M. A., 1995 (edit). Biological diversity in the drylands of the world. INCD, Geneve. 229 p.
Borodavka, D., 1988. Masters of their land. *Turkmenskaya Iskra*, 7 July: 3. (Russian)
Bovin, A., 1950. Green Belt of Karakum. *Pravda*, 16 September: 1. (Russian)
Brem., A., 1902. The life of animals. P. P. Soikin Publ., Saint Petersburg, 524 p. (Russian)
Bushev, A., 1992a. President of Turkmenistan eliminates the blunders of the Ministry of Water Management. *Komsomol'skaya Pravda*, 28 March: 3. (Russian)
Bushev, A., 1992b. Text books have got into a pretty mess. *Komsomol'skaya Pravda*, 12 December: 5. (Russian)
Butenina, V. & Gazman, V., 1991. Ecological order of the state. *Planovoe Khozyaistvo* 4:102-106. (Russian)
Bykov, I.,1880.Description of the Amydarya valley. Tashkent, 62 p. (Russian)
Cardy, F., 1994. Environment and forced migration. The Forth International Research and Advisory Panel Conference "New Directions in Research

and Forced Migration". Somerville College, University of Oxford, 5 - 8 January, Oxford, 16 p.
Central Asia in Figures, 1931. Sreazgosplan Publ., Moscow, 293 p.(Russian)
Central Asia. Physical environment and natural resources, 1968. Nauka Publ., Moscow, 483 p. (Russian)
Central Asia. Economico-geographical characteristics and the problems of development, 1970. Nauka Publ., Moscow, 502 p. (Russian)
Chaikovsky, A., 1884. Turkestan and its rivers as characterized by the Bible and Herodotus. Vladimir, 30 p. (Russian)
Charukhchev, N., 1988. We are excluding patrons. *Turkmenskaya Iskra*, 21 June: 3. (Russian)
Charykliev, Sh., 1988. Nomads against their will. *Turkmenskaya Iskra*, 14 July: 3. (Russian)
Churakova, E., 1998.Three millions people will come to Russia by the beginning of the 3rd millenium. *Finansovye Izvestiya*, 13 January: 6. (Russian)
Conquest, R., 1986. The harvest of sorrow: Soviet collectivization and the terror famine, London, 412 p.
Davis, K. P., 1966. Forest management: regulation and valuation. McGraw-Hill Book Co., New York, 519 p.
Davletov, D., 1977. Turkmen village in the end of 19 – beginning of 20th Centuries, Ashkhabad, 223 p. (Russian)
Dmitriev, I. D., Murakhtanov, E. S. & Sukhikh, V. I., 1989. Forest aerial photography and forest aviation. Agropromizdat Publ., Moscow, 366 p. (Russian)
Dobrin, L. G., 1964. The formation of barkhan. *Proc. Acad. Sci. Turkmenistan, Ser. Biol. Sci.*, 6 : 18 – 23. (Russian)
Donovan, E., 1883. A story of adventures and captivity. London, 348 p.
Down, K., 1997. The former republics of the USSR are more corrupted than other countries. *Finansovye Izvestiya*, 6 November 1997: 5. (Russian)
Dukhovnyi, V. A., 1980. Golodnaya Steppe – an example of complex development. *Problemy Osvoeniya Pustyn'* 6: 3 –16. (Russian)
Dukhovnyi, V. A., 1985. Give preference to economy. *Izvestiya*, 9 January: 5. (Russian)
Du Plessis, W. P., 1999. Linear regression relationships between NDVI, vegetation and rainfall in Etosha National Park, Namibia. *Journal of Arid Environment*, **42** (4): 236 – 260.
Durdyev, T., 1984. Victory of the cultural revolution in Turkmenistan, 1917-1941, p.130 – 132. TSSR. Encyclopaedia. Ashkhabad. (Russian)
Dyuzhev, G. A., 1984. Chemistry, p. 321 – 232. TSSR. Encyclopaedia. Ashkhabad. (Russian).
Economy of Commonwealth countries. Short statistical directory 1993. Finstatinprom Publ., Moscow, 384 p. (Russian)
Elizarov, Kh.,1980. How the pastures must be supplied with water. *Turkmenskaya Iskra*, 12 March: 3. (Russian)
Eyre, L. A., 1988. Population pressure on arid lands: is it manageable? Arid Land Today and Tomorrow, p. 989 – 996. Proc. Inter. Res. and Development Conf., Tucson, Arizona, USA, 20 – 25 October, 1985, London.
Fedin, V. P., 1983. About the biography of the Kara-Bogaz-Gol. *Nauki o Zemle* 10: 6-46. (Russian)
First general census of population in the Russian Empire, 1897. Saint Petersburg, 752 p. (Russian)
Fisak, V., 1988. The zone of special attention. *Turkmenskaya Iskra*, 20 August: 3. (Russian)
Flint, V. E. & Prisyazhnyuk V. E., 1986. Status of population, number and prospects of rehabilitation of dzheiran in the USSR. In: Protection and rehabilitation of dzheiran in the USSR. Gosagroprom Publ. Moscow. p. 3 – 9. (Russian)
Floret, Ch. & Potanier, R., 1982. L'ariditй en Tunisie presaharienne. Doc. No 150, ORS-TOM Publ., Paris, 544 p.
Gafurov, K. & Abdullaev, S., 1982. Soil cover of the irrigated zone of the Bukhara region. Tashkent, 130 p. (Russian)
Gapurov, M. G., 1982. The Soviet Turkmenistan. Politizdat Publ., Moscow, 128 p. (Russian)
Gapurov, M. G., Roslyakov, A. A. & Annanepesov, M., 1983. Fraternity for always. To the centenary of voluntary joining Turkmenistan to Russia. Ashkhabad, 89 p. (Russian)
Garbuk, S. V., & Gershenzson, V. E., 1997. Space systems of remote sensing the Earth. Moscow, 217 p. (Russian)
Gicheru P. T., 1996. Water erosion indicators. *Desertification Control Bulletin* 29: 13 – 19.
Gladkov, N. A. & Rustamov, A. K., 1975. Animals of the cultural landscapes. Mysl' Publ., Moscow, 219 p. (Russian)
Glantz, M., 1995. Sustainable development and accumulated ecological problems of the Aral Sea. National Center for Atmospheric Research. Boulder, Colorado, USA, 17 p.
Glazovsky, N. F. & Shestakov, A. S., 1995. Environmental migration caused by desertification. In : Desertification and migration. Geoforma Ediciones, Lograno, Espaта., p. 147–58.
Gorbunov, A. V. & Zarkhidze, V. A., 1986. Change in number, distribution and prospects of popula-

tion growth of dzheiran in Western Turkmenistan, p. 22 – 28. In: Protection and prospects of rehabilitation of dzheiran in the USSR. Gosagroprom Publ., Moscow. (Russian)

Grigor'ev, V. N.,1869. The Eastern and Chinese Turkestan. Saint Petersburg, 179 p. (Russian)

Grigor'eva, G.,1996. Monthly campaign on planting greenery. *Neitral'nyi Turkmenistan*, 15 March: 2. (Russian)

Grineva, G.,1992. The epoch has come of great geographical "closings". *Komsomol'skaya Pravda*, 30 October: 3. (Russian)

Grulev, M., 1901. The Amudarya. Description of the low flow. Tashkent, 92 p. (Russian)

Gunin, P. D., 1990. Ecology of desertification processes in arid ecosystems.VASKHNIL Publ., Moscow, 345 p. (Russian)

Gusseinov, E., 1996. The crown in outlying district of the former empire. *Izvestiya*, 19 September: 4. (Russian)

Il'khanov, A., 1992. Ethnic and social problems of the transition period in Uzbekistan. *Sotsiologicheskie Issledovaniya* 8: 12 – 17. (Russian)

Illesh, A., 1985. When will the ghosts disappear? *Izvestiya*, 13 June: 5. (Russian).

Illesh, A. & Surkov, V., 1985. Paper ghosts from the "Tiger Ravine". *Izvestiya*, 5 January: 4. (Russian)

Il'yasov, F. & Mukhametberdiev, O.,1988. Will the hand become scare? *Turkmenskaya Iskra*, 26 June: 4. (Russian)

Inquest of population about the environmental quality and efficiency of measures on conservation of environment, 1991. Goskomstat Publ., Moscow, 130 p. (Russian)

Iqbaluddin, Saifuddin, Akram Javed & M. Monti, 1999. Terrain mapping unit-based approach for sustainable agricultural development in India – an example from plains ecosystem in semi-arid interface. *Journal of Arid Environments* 42(4): 320 – 330.

Isaev, A. S., 1991 (Ed.). Prognosis of exploitation and regeneration of forest resources in economic regions of the USSR till 2000, Vol. **II**, Moscow, 993 p. (Russian)

Ivanin, M. I.,1873. Khiva and the Amudarya river. Saint Peteraburg, 24 p. (Russian)

Ivanov, A. P., 1972. Physical principles of the deflation in the desert. Ylym Publ. Ashkhabad, 117 p. (Russian)

Ivanov, N. P., 1938. The order of Nadir Shakh to his warriors from Merv on restoration of the town of Merv, p. 150 – 157. In: Materials on history of Turkmen and Turkmenia. AN SSSR Publ., Moscow. (Russian)

Ivchenko, M., 1992. Correcting the mistakes of many years. *Izvestiya*, 21 October: 4 (Russian)

Jenkins, G., 1974. A note on climate cycles and the rise of Chinggis Khan. *Central Asian Journal*, 18: 17-31.

Kadymov, G., 1993. Why Middle Asia has become Central Asia? *Turkmenistan Segodnya*, 4 March: 1. (Russian)

Kakhkharov, A. G., 1993. Socio-ecological situation in Uzbekistan. *Sotsiologicheskie Issledovaniya*, 1: 119 – 123. (Russian)

Kalenov, G. S., 1992. Ecological and geographical differences in desert vegetation. Doctor's thesis. Desert Research Institute, Ashkhabad. 49 p. (Russian).

Karpov, A., 1990. The sentence has passed. *Izvestiya*, 6 February: 4. (Russian)

Karpov, V., 1988. Suffering in Khauzkhan. *Turkmenskaya Iskra*, 5 July : 4. (Russian)

Kasymova, E., 1995. If we have forage we shall have meat. *Turkmenskaya Iskra*, 28 February: 2. (Russian)

Keith, H. G., 1965. Libyan flora. United Middle East Executives Ltd., Vol. 1 – 2, London, 1047 p.

Kerimov, D., 1988. The volume of forage will increase. *Turkmenskaya Iskra*, 23 June: 3. (Russian)

Kes', A. S., 1939. The river bed of the Uzboi. Proc. Institute of Geography, Vol. XXX. AN SSSR Publ., Moscow,120 p. (Russian)

Khabibulaev, A. M., 1993.Ecological problems in the Aral Sea basin and how to solve them. *Problemy Osvoeniya Pustyn'* 3: 24 – 28. (Russian)

Khanykov, N., 1843. Description of the Bukhara Khanate. Saint Petersburg, 210 p. (Russian)

Kharin, N.G., 1975. Remote sensing the vegetation. Nauka Publ., Moscow, 130 p. (Russian)

Kharin, N.G., 1994. Desertification of the arid lands of Turkmenistan, p. 65 – 76. In: Biogeography and Ecology of Turkmenistan. Kluwer Academic Publ., London.

Kharin, N.G., 1995. Change of biodiversity in ecosystems of Central Asia under the impact of desertification, p. 23 – 32. In: Biodeversity in the drylands of the world. INCD, Geneve.

Kharin, N.G., 1996a. Strategy to combat desertification in Central Asia. *Desertification Control Bulletin*, 29 : 29 – 34.

Kharin, N.G.1996b. Water scarcity in Central Asia. In: Water Resources Management and Desertification: Problems and Challenges. WMO, Geneve.

Kharin, N.G., 1998. Estimation of desertification damage and the cost of land rehabilitation: A methodological approach. *Desertification Control Bulletin* 33 : 29 – 32.

Kharin, N.G. & Kalenov, G.S., 1978. Study of human-induced desertification by space photos. *Problemy Osvoeniya Pustyn'* 4: 25 – 27. (Russian)

Kharin, N.G., Kalenov, G.S. & Kiril'tseva, A.A. et al., 1988. Explanatory note to the map of man-made desertification of the USSR arid lands, scale 1: 2,500, 000. Ashkhabad, 29p. (Russian)

Kharin, N.G., Kalenov, G.S. & Kiril'tseva, A.A. et.al., 1993. Explanatory note to the map of human-induced land degradation in the Aral Sea basin, scale 1: 2,500,000. Ashkhabad, Desert Research Institute, 83p.

Kharin, N.G., Kalenov, G.S. & Kurbanmuradov, K. et al., 1978. Methodology of application of space photos for compilation and correction of thematic maps in the desert zone. Ashkhabd, Desert Research Institute, 75p. (Russian)

Kharin, N.G., Kalenov, G. S. & Volovik, V. V., 1993. Map of drylands of Asia. Desert Research Institute, Ashkhabad.

Kharin, N.G., Nechaeva, N.T. & Nikolaev, V. N. et al., 1985. Methodological principles of desertification assessment and mapping. Desert Research Institute, Ashkhabad, 117 p.

Kharin, N.G, Nikolaev, V.N., Sorokina, I.P., & Kiril'tseva A. A, 1989. About modelling of desertification processes using natural physical and anthropogenic factors.*Problemy Osvoeniya Pustyn'* 3 : 7 – 12. (Russian)

Kharin, N.G., Nurberdyev, K. & Serkhenov, E., 1969. Use of aerial photography for assessment and mapping sands. *Vestnik Sel'sko-khozyaistvennoi Nauki* 6: 74 - 80. (Russian)

Kharin, N.G., Orlovsky, N.S. & Alferov, G.A., 1990. Desertification assessment and mapping in Mali. *Arid Lands Newsletter, Fall, Winter*, 30 : 30 – 32.

Kharin, N.G. & Petrov, M. P., 1975. Glossary of terms on desert environment and land reclamation. Ashkhabad, 141 p.

Kharin, N.G. & Pratov, U., 1982. Meeting with Silphium after 2,000 years. *Priroda* 4: 92 – 96. (Russian)

Kharin, N.G., Tateishi, R. & Gringof I.G., 1998. Use of NOAA/ AVHRR data for assessment of ptecipitation and land degradation in Central Asia. *Arid Ecosystems* 4 (8) : 25 – 34.

Kharin, N. G., Tateishi, R. & Harahsheh, H., 1999. Degradation of the drylands of Asia, 81 p.+ Desertification map of the drylands of Asia, scale 1:10,000,000. CEReS, Chiba University, Japan.

Khodzhamuradov, A., 1984. Water management and melioration, p. 226 – 228. TSSR. Encyclopaedia, Ashkhabad. (Russian)

Khudaikuliev, A. G., 1984. Cotton growing, p. 217 – 218. TSSR. Encyclopaedia, Ashkhabad. (Russian)

Khudaikuliev, M.,1984. Forestry, p. 230. TSSR. Encyclopaedia, Ashkhabad. (Russian)

Khudainazarov, B.,1988. Traces in the sand. *Turkmenskaya Iskra*, 1 April: 4. (Russian)

Kidenis, A.,1997. Tajik way to market. *Izvestiya*, 20 November: 4. (Russian)

Kiril'tseva, A. A., 1986. Optical indicators of desertification. *Problemy Osvoeniya Pustyn'* 4: 38 – 41. (Russian)

Kirsta, B. T., 1991. Resources of surface water and natural humidity of Turkmenistan territory. Ilym Publ., Ashkhabad, 179 p. (Russian)

Kiyatkhanov, B. A. & Tesler, D. V., 1984. The progress in public health service in Soviet Turkmenistan, p. 256 – 261. TSSR. Encyclopaedia. Ashkhabad. (Russian)

Kogan, Sh. I., 1991. The Karakum Canal and its life. Ylym Publ., Ashkhabad, 75 p. (Russian)

Kolodin, M.V., 1992. Ecological and economic problems of water conservation in the arid zone. Ylym Publ., Ashkhabd, 230 p. (Russian)

Kolyabin, V., 1988. The contrasts of Khauz-Khan. *Turkmenskaya Iskra*, 19 April:3 (Russian)

Kondrashov, S., 1994. Where the evel of day is combined with eternity. *Izvestiya*, 10 June: 4. (Russian)

Köppen, W., 1918. Klassifikation der Klima nach Temperatur. Niederschlage und Jahreslauf. *Petermanns Mitteilungen* **64**: 193-203.

Korovin, E.P., 1961. Vegetation of Central Asia. Part I, Tashkent, 452 p. (Russian)

Korzh, A., 1990. Formation of melioration properties of irrigated soils in the Tashauz oasis. *Melioratsia i Vodnoe Khozyaistvo* 2: 8 – 11. (Russian)

Kovalev, A., 1990. All people must know all truth. *Izvestiya*, 8 February: 4. (Russian)

Kovalevsky, S.A., 1924. The volcano Bukhty and tectonics of Bibi-Aibat. *Azerbaidzhanskoe Neftyanoe Khozyaistvo* 23: 50-65. (Russian)

Kovalevsky, V., 1985. What was hidden behind the replies? *Izvestiya*, 24 September: 5. (Russian)

Kovda, V.A., 1950a. The plan of the taming the desert. *Pravda*, 13 September: 1. (Russian)

Kovda, V. A., 1950b. The results and new tasks of soil study in Middle Asia, p. 44 – 45. In: Deserts of the USSR and their development. AN SSSR Publ., Moscow. (Russian)

Kovda, V. A.,1977. Aridization of drylands and drought control. Nauka Publ., Moscow, 272p. (Russian)

Kovda, V. A., 1989. Problems of soil conservation and the biosphere protection. Pushchino, 155p. (Russian).

Krutov, A. N., 1995. Environmental degradation of the Aral Sea zone and project on stabilization of the ecological situation. *Problemy Osvoeniya Pustyn'* 6 : 17- 24. (Russian)

Kuleshov, V., 1988a. Cast-iron fences remain impregnable. *Izvestiya,* 13 March: 4. (Russian).

Kuleshov, V., 1988b. Technical means of atheism. *Izvestiya*, 14 July: 3. (Russian)

Kuleshov, V., 1990a. Ex manager has condemned for bribes. *Izvestiya*, 9 February: 3. (Russian)

Kuleshov, V., 1990b.Drug business band in the dock. *Izvestiya*, 11 February: 2. (Russian)

Kuleshov, V., 1992. Cobra and gyurza are staying under the protection of president. *Izvestiya*, 21 July : 4. (Russian)

Kuleshov, V., 1993. Fines for poaching have been increased 10 times, but animals are shot in reserves. *Izvestiya*, 19 May : 7. (Russian)

Kuliev, N., 1988. Technical means of atheism. *Turkmenskaya Iskra*, 14 July: 3. (Russian)

Kurbanov, E., 1984. Raising sheep and goats, p. 223. TSSR. Encyclopaedia. Ashkhabad. (Russian)

Kurbanov, O., 1979. Hunting and poaching. *Turkmenskaya Iskra*, 13 June: 2. (Russian)

Kurbanov, O., 1985. The roads, the roads... *Vecherny Ashkhabad*, 23 September: 3. (Russian)

Kuropatkin, A. N., 1899. The conquest of Turkmenia. Saint Petersburg, 224 p. (Russian)

Latipova, L., 1993. Cholera was registered, but the information about that should be paid. *Izvestiya*, 13 August: 2. (Russian)

Lavrov, A. P., 1964. Experimental assessment of the degree of sand deflation. *Proc. Turkmenistan Ac. Sci. (Biological Series)* 1: 12 - 14. (Russian)

Lavrov, A. P., 1984. The soils, p. 45 - 51. TSSR. Encyclopaedia. Ashkhabad. (Russian)

Le Houerou, H. N., 1992. Relation entre la variabilité de precipitation et celle des production primaire et secondaire en zone aride. In: L'aridité: une contrainte au developpement, p. 198 - 220. Edition de l'ORSTROM, Paris.

Le Houerou, H. N., 1994. Zero-order draft on climate change, drought and desertification. Intergovernmental Panel on Cimate Change, Geneva, 22p.

Leshchinsky, G. T., 1974. Resources of ephemeral runoff in the deserts of Central Asia and Western Kazakhstan. Ylym Publ., Ashkhabad, 186 p. (Russian)

Leskov, S., 1993. Lavrentii Beriya could achieve the prosperity of the country. *Izvestiya*, 29 January: 4. (Russian)

Lessar, P. M., 1884. Notes about the transcaspian region and bordering countries. Saint Petersburg, 169 p. (Russian)

Le Vein, S., 1997. Sweet promises in Uzbekistan have a sour taste. *Izvestiya*, 22 May: 3. (Russian)

Lobova, E. V., 1971. Soil map of Central Asian Republics, scale 1 : 2, 500, 000. GUGK SSSR Publ., Moscow.

Loginov, V., 1989. Even the arm-chair has been not unsteady or why the inerrables don't loose the force. *Turkmenskaya Iskra*, 1 April: 3. (Russian)

Luk'yanenko, P., 1993. Ivanov has disturbed somebody. *Argumenty i Fakty*, September, 35: 2 (Russian).

Lynev, R., 1988. From the thaw to the stagnation. *Izvestiya*, 18 November: 4. (Russian)

Mainguet, M., 1996. Aridité, sécheress et dégradation dans les aires séche de Chine. *Sécheresse*, 7: 41-50.

Malikov, A., 1996. Travel from Tajikistan to Russia with wodka. *Argumenty i Fakty*, December 27:10. (Russian)

Mamedov, E. D., 1984. On pluvial paleoclimates of deserts in Pleistocene. *Problemy Osvoeniya Pustyn'* 3: 10 –16. (Russian)

Mamedov, V., 1988. Non-inborn defects. *Turkmenskaya Iskra*, 8 June: 2. (Russian)

Masson, V. M., 1966a. Agricultural neolith in the southwestern regions of Central Asia, p. 76 – 92. In: Central Asia in epoch of stone and bronze. Nauka Publ., Moscow. (Russian)

Masson, V. M., 1966b. Flourishing and decay of ancient agriculture in the southwestern regions of Central Asia, p. 151 – 178. In: Central Asia in epoch of stone and bronze. Nauka Publ., Moscow. (Russian)

Masson, V. M., 1971. The settlement of Dzheitun. Nauka Publ., Moscow. 208 p. (Russian)

Masson, V. M., 1977. Desert and society: the dynamics of the ir interaction during the historical period. Problemy Osvoeniya Pustyn' 6: 3 – 10. (Russian)

Materials of the 25th Congress of the KPSS, 1976. Politizdat Publ., Moscow, 203p. (Russian)

Meigs, P., 1951. Distribution of arid homoclimates. UNESCO Publ., Paris, 19p.

Mez, A.,1922. Die Renaissaance des Islams. Heidelberg, 512p.

Mikhailov, V. & Smol'nikov, G., 1999. Niyazov has commanded to plant trees. *Nezavisimaya Gazeta*, 8 September: 3. (Russian)

Mikheev, V., 1996. Refugees of Russia are taken under the protection of UN. *Izvestiya*, 16 October: 3. (Russian)

Mlechkin, L., 1993. "Lesha Gorbun" is ruling over Pamir. *I zvestiya*, 17 December: 3. (Russian)

Mollaeva, A., 1984. People's education, p. 267 – 282. TSSR. Encyclopaedia. Ashkhabad. (Russian)

Morozova, G. F., 1992. Up-to-date migration process: refugees and emigrants. *Sotciologicheskie Issledovaniya* 3: 34-40. (Russian)

Moser, H., 1885. A travers l'Asie Central. Paris, 458p.

Mostovshchikov, S., 1991. Gdlyan, Ivanov and Kalugin are considered to be decent people. *Izvestiya*, 4 September: 4. (Russian)

Mukasheva, Zh. N., 1992. Landscapes of the Low Amudarya and their vulnerability to human impact. Summary of Thesis. Desert Research Institute, Ashkhabad, 18 p. (Russian)

Mukhammedov, G. M.,1979. Improvemnt of rangelands in Central Karakum. Ylym Publ., Ashkhabad, 215 p. (Russian)

Mukhammedov, N., 1988. Ideological falsification. *Turkmenskaya Iskra*, 26 June: 3. (Russian)

Mukhammetberdiev, O.,1988. Once more about empty buckets and superstition. *Turkmenskaya Iskra*, 14 May: 4. (Russian)

Murkina, N. V., 1976. Ecology of *Rana nigromaculata* in the Karakum. In: Theoretical and practical aspects of hunting, p. 120 – 133. Minsel'khoz SSSR Publ., Moscow. (Russian)

Musienko, E., 1994. End of great chase. *Subbota*, 26 February: 7. (Russian).

Myalo, K. & Goncharov, P., 1989. The glow of Fergana. Novoe Vremya 38: 31 – 33. (Russian)

Nan-Ting Chon & Dregne, H. E., 1993. Desertification control: cost and benefit. *Desertification Control Bulletin* 22: 20 – 26.

Nadein, V., 1985. Laurel kick. *Izvestiya*, 11 January: 4. (Russian)

National action program to combat desertification in Turkmenistan, 1996. Desert Research Institute/UNEP. Ashkhabad, 106p.

National economy of the USSR, 1972. Statistika Publ., Moscow, 847p. (Russian)

National economy of the USSR, 1980. Statistika Publ., Moscow, 615 p. (Russian)

National economy of the USSR, 1991. Statistika Publ., Moscow, 820 p. (Russian)

Nechaeva, N. T., 1953. Sheep raising on rangelands of Turkmenistan. AN TSSR Publ., Ashkhabad, 70 p. (Russian)

Necheva, N. T. & Nikolaev, V. N., 1962. Explanatory note to the rangeland map of plain Turkmeniya, scale 1: 1,000,000. Ylum Publ., Ashkhabad, 125 p. (Russian)

Nechaeva, N. T. & Prikhod'ko, S. Ya., 1966. Artificial pastures in foothill deserts of Central Asia. Ylym Publ., Ashkhabad, 224 p. (Russian)

Nechaeva, N. T., Prikhod'ko, S. Ya., Bashkatova, A. N. & Kayanova, R. M., 1959. Experimental melioration of desert rangelands in Turkmenistan. AT TSSR Publ., Ashkhabad, 175 p. (Russian)

Nechaeva, N.T., Vasilevskaya, V. K. & Antonova, K.G., 1973. Life forms of plants in the Karakum desert. Nauka Publ., Moscow, 241 p. (Russian)

Nechaeva, N. T., Vasilevskay, V. K., & Antonova, K. G., 1979. Productivity of plants in Central Karakum in connection with different regime of utilization. Nauka Publ., Moscow, 254 p.(Russian)

Neronov, V. M. & Bobrov, V. V., 1990. Conservation of rare animals in the deserts of the USSR. *Arid Land Newsletter. Fall/ Winter* 30: 15-19.

Nikolaev, V. N.,1972. Natural forage resources of Turkmenistan. Ylym Publ., Ashkhabad, 215p. (Russian)

Nishanov, R. N., 1988a. The speech of comrade Nishanov. *Pravda*, 1 June: 2. (Russian)

Nishanov, R. N., 1988b. The speech of deputy Nishanov. *Izvestiya*, 30 November: 3. (Russian)

Niyazov, S. A., 1988. The speech of comrade Niyazov. *Turkmenskaya Iskra*, 24 May: 1. (Russian)

Nosko, V., 1988. Contrasts in the farms. *Turkmenskaya Iskra*, 6 April: 4. (Russian)

Nurberdyev, K., 1982. Soil erosion on the piedmont plain of the Kopetdag and techniques to control it. Ylym Publ., Ashkhabad, 200 p. (Russian)

Nurgel'dyev, O. N., 1969. Ecology of mammals in the plain Turkmenia. Ylym Publ., Ashkahabd, 230 p. (Russian)

Obramenko, B., 1988. Under the cover of paper leaves. *Turkmenskaya Iskra*, 21 August: 3. (Russian)

Ochakovskaya, L., 1985. We are ready to meet spring. *Izvestiya*, 2 March: 3. (Russian)

Okladnikov, A. P., 1966. Middle Asia in Quarternary time. Relics of the low Paleolith, p. 11 – 22. In : Middle Asia in epoch of stone and bronze. Nauka Publ., Moscow. (Russian)

Oldeman L. R., Hakkeling R. T. A., Sombroek W.G., 1991. World map of the status of human-induced

soil degradation. An explanatory note. UNEP, 34 p.
Orlov, M. M., 1927. Forest management. Lesprom Publ., Leningrad. 213 p. (Russian)
Orlovsky, N. S., 1977. Effect of irrigation and watering on desert microclimate. *Problemy Osvoeniya Pustyn'* 5: 23-27. (Russian)
Osipov, G., 1988. General care of harvesting. *Turkmenskaya Iskra*, 30 September: 3. (Russian)
Otterman, J., 1974. Baring high-albedo soils by overgrazing. Science **86** : 717-721.
Ovcharenko, G. & Chernenko, A., 1988. The son in law and his friends. *Pravda*, 30 August: 3. (Russian)
Ovezov, B., 1967. The Karakum Canal and development of economy in Turkmenistan. Ylym Publ., Ashkhabad, 237 p. (Russian)
Palen, K. K., 1910a. Irrigation in Turkestan. Saint Petersburg, 478 p. (Russian)
Palen, K. K., 1910b. Report on the inspection of Turkestan territoty. Saint Petersburg, 263 p. (Russian)
Palen, K. K., 1911. About characteristics of the national economy in Turkestan. Part II. The tables. Saint Petersburg, 262 p. (Russian)
Pavlov, V. A.,1994. Seeds from the Netherlands, potato from Iran. *Vecherny Ashkhabad*, 23 February: 2. (Russian)
Pavlova, A., 1990. I allow the bee to take off. *Izvestiya*, 11 November: 4. (Russian)
Petrov, M. P., 1973. Deserts of the world. Nauka Publ., Moscow, 433 p. (Russian)
Petrov, M. P., 1975. About the limits of arid regions. *Problemy Osvoeniya Pustyn'* 2 : 3-10. (Russian)
Pilyatsky,B., 2000. Ecological crimes could be punished by capital punishment. *Izvestiya*, 12 February: 5. (Russian)
Pluzhnik, A., 1994. Wax figures are living on the stage of history. *Izvestiya*, 28 February: 5. (Russian)
Polovinkina, L., 1988a. We shall not permit to harm him. *Turkmenskaya Iskra*, 5 July: 4. (Russian)
Polovinkina, L., 1988b. What could do women council. *Turkmenskaya Iskra*, 9 August: 4. (Russian)
Pral'nikov, A., 1985. Where saigak is running. *Izvestiya*, 11, January: 4. (Russian).
Prince, S.D., & Astle, W. L., 1986. Satellite remote sensing of rangelands in Botswana. 1. Landsat MSS and herbaceous vegetation. *Int. J. Remote* Sensing **I 7** (11): 1534 – 1554.
Prokhorov, V., 1983. Don't sleep marmot! *Pravda*, 25 June: 4. (Russian)
Prokhvatilov, V., 1992. How can we arrange the life in CIS. *Komsomil'skaya Pravda*, 26 September : 5. (Russian)
Provisional methodology for desertification assessment and mapping,1981.FAO/UNEP, Rome, 57p.
Provisional methodology for assessment and mapping desertification, 1983. FAO/UNEP, Rome, 104p.
Putorkina, A.,1988. Double face.*Turkmenskaya Iskra*, 12 July: 4. (Russian)
Questening the population. Sociological study, 1991. Finstatinform Publ., Moscow, 412 p.
Rafikov, A. A., 1984. Assessment of meliorative properties of irrigated soils in the southern part of the Aral Sea zone. Tashkent, 160 p. (Russian)
Rakhmanov, Zh. V., 1994. Analysis of up-to-date exogenous processes in Govurdak – Koiten region. Summary of Thesis. Desert Research Institute, Ashkhabad, 17p.
Raskin, P., Khansen, E., Zhu, Zh. & Stavitsky, D., 1993. Modelling of water resources and their consumption in the Aral Sea region. *Problemy Osvoeniya Pustyn'* 3: 28 – 43. (Russian)
Radziminsky, P. Z.,1975. Some data on spectral reflectance of plants in the desert zone. *Problemy Osvoeniya Pustyn'* 1: 12 – 21. (Russian)
Razakov, P. M., 1990. Ecological measures in the Aral bordering areas: Investigation and the program of action. Melioratsiya i Vodnoe Khozyaistvo 1: 13 – 18. (Russian)
Razgulyaev, Yu., 1996. Gangsters were not condemned by public opinion. *Pravda*, 26 July: 6. (Russian)
Red Data Book of the USSR, 1978. Lesnaya Promyshlennost' Publ., Moscow, 459 p. (Russian)
Red Data Book of Turkmen SSR, 1985. Turkmenistan Publ., Ashkhabad, 403 p. (Russian)
Reklyu, E., 1892. The land and the peoples. Vol. VI. Saint Petersburg, 700 p. (Russian)
Rickmers, W. R., 1913. The Duab of Turkestan. A physical sketch and account of some travel. Cambridge, 546 p.
Rustamov, A. K., 1976. The role of water in the desert from the point of view of ecology and zoogeography, p. 40 – 44. In: Theoretical and practical aspects of nature conservation and hunting, p. 40 – 44. Minsel'khoz SSSR Publ., Moscow. (Russian)
Rustamov, A. K., 1986. Dzheiran in Turkmenistan: the status of population, distribution, breeding and protection. In: Conservation and prospects of rehabilitation of dzheiran in the USSR. Gosagroprom SSSR Publ., Moscow. (Russian)

Rustamov, A. K., 1991. Some problems of biodiversity conservation under the conditions of human-induced desertification. *Problemy Osvoeniya Pustyn'* 3 – 4: 31 – 36. (Russian)

Ryabova, P., 1988. The teacher is a personality. *Turkmenskaya Iskra*, 1 April: 3. (Russian)

Ryabushkin, A., 1993. From Kirghizstan abroad to find a job. *Izvestiya*, 13 January: 4. (Russian)

Sabilaev, A. S. & Koptsov, L. A., 1986. The status and the past of dzheiran population in the Karakalpakian Ustyurt, p. 12 – 16. In: Conservation and prospect of rehabilitation of dzheiran in the USSR. Gosagroprom USSR Publ., Moscow. (Russian)

Safarov, I.S., & Safarov, N. I., 1995. The importance of scientific prognosis. Elm Publ., Baku, 32 p. (Russian)

Sadykov, K.S., 1988. Speech of deputy Sadykov. *Izvestiya,* 30 November: 3. (Russian)

Sakhnin, A., 1988. Corruption. *Izvestiya*, 2 November: 3. (Russian)

Seiitkurbanov, S., 1991. The program of utilization of natural resources and nature conservation in Turkmenistan till 1991 – 1995 and further till 2000. Ashkhabad, 65p. (Russian)

Savinykh, V.P., Kuchko, A.S. & Stetsenko, A.F., 1997. Aerial and space photography. Geodezizdat Publ., Moscow, 378 p. (Russian)

Semenov-Tianshansky, V. P., 1913. Full geographical description of our mother country. Vol. 19. Turkestan region. Saint Petersburg, 861 p. (Russian).

Sheripov, M., 1984. Pig breeding, p. 224. TSSR. Encyclopaedia. Ashkhabad. (Russian)

Shul'ts, V. L., 1985. Rivers of Central Asia. Gidrometeoizdat Publ., Leningrad, 238 p. (Russian)

Singh, S., Kar, A. & Joshi, D. S. et al.,1992. Desertification map of Rajastan. *Annals of Arid Zones* **34** (4): 17 – 23.

Skobelev, M. D.,1881. The siege and capture of the fortress Dzhengil' Tepe or Geok Tepe. Saint Petersburg, 64 p. (Russian)

Slutnik, V., 1988. The time for persistent work. *Turkmenskaya Iskra*, 20 May: 2. (Russian)

Smirnov, K. I., 1993. Poacher is the most terrible beast of pray. *Izvestiya*, 7 April: 4. (Russian)

Soderling-Brydolf Ch.,1973. The flower from Cyrena. Nauka Publ., Moskow, 118p. (Russian).

Sokolov, A. I., 1986. Distribution, number and status of dzheiran population in Tajikistan, p. 37 – 39. In: Conservation and prospects of rehabilitation of dzheiran in the USSR. Gosagroprom Publ., Moscow. (Russian)

Sokolov, V. E., 1990. The Badkhyz Reserve, p. 198 - 213. In: The Reserves of Central Asia and Kazakhstan. Mysl' Publ., Moscow. (Russian)

Stamp L.D., 1961. A glossary of geographical terms. Vol. 2, London, 385 p.

Stiles,D., 1995. Linkage between dryland degradation and migration: a methodology. *Desertification Control Bulletin*, 30: 9 –18.

Sukhikh, V. I., 1991. Problems of forestry in Central Asia and Kazakhstan. *Lesnoe Khozyaistvo* 6: 2 – 7. (Russian)

Sukhikh, V. I., Kharin, N.G. & Butusov, O.B., 1999. Preliminary north taiga forest cover classification with satellite "Resource-1" data. *Investigation of Earth from Space* 5: 65 – 74. (Russian)

Suleimanova, D., 1988. School compositions. *Turkmenskaya Iskra*, 21 May: 4. (Russian)

Sultangazin, I.M., 1991. An concept of conservation and rehabilitation of the Aral Sea, in particular – normalization of ecological and socio-economic conditions in the Aral Sea region. *Problemy Osvoeniya Pustyn'* 3 – 4: 97 – 107. (Russian)

Svintsov, I. P., 1988. Melioration if sandy deserts in Turkmenistan. Ylym Publ., Ashkhabad, 159p. (Russian)

Tachmamedov, K., 1989. To eradicate misfortune. *Turkmenskaya Iskra*, 18 April: 3 (Russian).

Tashliev, A. O., 1988. Resources of fauna in Turkmenistan and the problems of their rational utilization. Ylym Publ., Ashkhabad, 102 p. (Russian)

Tateishi, R.& Wen Cheng Gang, 1995. Land cover database for Asia. Proc. Int. Symposium on Vegetation Monitoring, August 29 – 31, Chiba University, Japan, p. 13 – 19.

Technical rules for projecting and constructing the road-beds in the sandy deserts,1963. Dept. Transport Constructions, Moscow, 46p. (Russian)

Terent'ev, M. A., 1906. The history of conquest of Central Asia. Vol. II, Saint Petersburg, 318p. (Russian)

The Main Turkmen Canal, 1952. AN SSSR Publ., Moscow, 307 p. (Russian)

Tolba, M., 1985. Heads in the sand. A new appraisal of arid land management. UNEP, Nairobi, 12p.

Toshchenko, Zh. T., 1992. Russian refugees: a tragedy or the cost of the imperial ideology. *Sotsiologicheskie Issledovaniya* 9: 59 – 64. (Russian)

Townshed, J. R. G., 1994. Global data sets for land study application from Advanced Very High

Resolution Radiometer. *Int. J. Remote Sensing* **15** (17): 3319 – 3332.

Tucker, C., Sellers., P. J., 1985. Satellite remote sensing of primary production. *Int. J. Remote Sensing* (**7**)11: 1395 – 1416.

Turkmenbashi, S., 1995. The care about people should be not stopped. Turkmenskaya Iskra, 7 February: 1. (Russian)

Tursunov, L.T., 1981. Soil properties of the irrigation zone in western Uzbekistan. Tashkent, 223 p. (Russian)

Ujfalvy, M., 1880. De Paris a Samarkand. Paris, 360 p.

United Nations Convention to Combat Desertification in Those Countries Experiencing Serious Drought and/or Desertification, Particularly in Africa, 1994. UNEP, Geneva, 71 p.

Uraev, B., 1995. Fishing began in Sarykamysh Lake. *Turkmenskaya Iskra*, 20 October: 2. (Russian)

Urazaev, Sh. A., 1958. The Turkestan ASSR and its state and lawful status. Tashkent, 217p. (Russian)

Ustaev, A., 1995a. Carelessness is punishable. *Turkmenskaya Iskra*, 25 October: 3. (Russian)

Ustaev, A, 1995b. Poachers are arrested. *Turkmenskaya Iskra*, 31 January: 3. (Russian)

Vagabov, M. V., 1968. Islam and woman. Mysl' Publ., Moscow, 229 p. (Russian)

Vambery, A.,1865. Voyage d'un faux dervich. Paris, 212p.

Vasil'ev, Yu., 1991. The trial on Adylov will take place soon. *Izvestiya*, 13 March: 5. (Russian)

Vinogradov, B., 1996. Successors of Timur go on their own path. *Izvestiya*, 1 October: 4. (Russian)

Vladimirov, V., 1993a. The three, the six and the ace of hearts. *Argumenty i Fakty*, 2 January: 10. (Russian)

Vladimirov, V., 1993b. Invocation of marshal Grechko. *Argumenty i Facty*, 6 February: 7. (Russian)

Volovik, V. V., 1992. The main stages of anthropogenic changes of landscapes in the Northern Turkmenistan. *Problemy Osvoeniya Pus-tyn'* 1: 27– 34. (Russian)

Voropaev, G.V., 1982. Water resources of Central Asia and the problem of their intensive utilization. *Problemy Osvoeniya Pustyn'* 1: 13 – 20. (Russian)

Voznesensky, A. N., 1972. Water, its resources and consumption. *Wodnye Ressursy* 1: 3 – 14. (Russian)

Watanaba, K., 1995. The world refugee situation and the environment. In: Desertification and migration. Int. Symposium on desertification and migration, 9 -11 February 1994, p. 295 – 317. Roquetas de Mar Almeriap. Lograno, Espата.

World Atlas of Desertification, 1992. UNEP, London, 69p.

World Bank, 1995. Project of improvement of water supply and drainage in Turkmenistan. Summary of the report, Washington, September, 14p. (Russian)

World Health Organization, 1994. Basic document on the aspect of human health in the Aral Sea region. Int Conference on Sustainable Development, Nukus, Uzbekistan, 18 – 20 September, 9p.

Yosikiro Takano, 1993. Ecological rehabilitation of the environment in the Aral Sea basin and joint international activities. *Problemy Osvoeniya Pustyn'* 3: 11- 16. (Russian)

Yuferev, V. I., 1925. The directory of cotton growing in the USSR. Moscow, 604 p. (Russian)

Yusarov, Kh., 1995. Paleography of the Uzboi based on historical sources. *Problemy Osvoeniya Pustyn'* 2: 45 – 53. (Russian)

Yusupov, U,. 1950. The great Stalin's building. *Pravda*, 15 September: 1. (Russian)

Zadneprovsky, Yu. A., 1966. The ancient Chust culture in the Valley of Fergana, p.193 – 207. In: Central Asia in epoch of stone and bronze, p.193 – 207. Nauka Publ., Moscow. (Russian)

Zaletaev, V. S., 1989. Ecologically destabilized environment. Ecosystems of arid zoneby changing hydrological regime. Nauka Publ., Moscow, 146p. (Russian)

Zha, Y. & Gao, J., 1997. Characteristics of desertification and its rehabilitation in China. *Journal of Arid Environments* 37 (3): 419 – 432.

Zakharchenko, V. T., 1984. The Lenin Karakum Canal, p. 228 – 230. TSSR. Encyclopaedia. Ashkhabad. (Russian)

Zhbanov, E., 1990. Why he left? *Izvestiya*, 23 March: 4. (Russian)

Zholdaev, A., 1990. Ecology... and ethnos. *Zvezda Vostoka* 12: 9 – 16. (Russian)

Ziyatdulaev, S.K., 1979. Social and economic problems of the partial transfer of the Siberian rivers flow to Middle Asia and Kazakhstan. FAN Publ., Tashkent. 61p. (Russian)

Zonn, I.S., Nikolaen, V.N. & Orlovsky, N.S., 1981. The practical experience in desertification control in the USSR. Nauka Publ., Mos-cow,114p. (Russian)

Zwemer, A. E. & Zwemer, S. W., 1926. Moslem women. Central Commission on Study of Foreign Missions. North Cambridge, Mass., USA, 272p.

INDEX
(pages of manuscript)

MIX
Papier aus verantwortungsvollen Quellen
Paper from responsible sources
FSC® C105338

If you have any concerns about our products,
you can contact us on
ProductSafety@springernature.com

In case Publisher is established outside the EU,
the *EU authorized representative is:*
Springer Nature Customer Service Center GmbH
Europaplatz 3, 69115 Heidelberg, Germany

Printed by Libri Plureos GmbH
in Hamburg, Germany